SOIL FERTILITY AND NUTRIENT MANAGEMENT

[A Way to Sustainable Agriculture]

The book entitled "Soil fertility and Nutrient management" is a compilation work and most of the information was farmed very critically covering all the main topics of plant nutrition. The book will be serve as useful reference to students, teachers, researchers scientists, policy makers and other interested in soil science, agronomy, crop science, environmental sciences and agriculture.

Mr. A. S. Jadeja: He was obtained B.Sc. (Hons.) Agri. and M.Sc. (Agri.) in Agril. Chem. & Soil Science degrees from Junagadh Agricultural University in 2014 and 2016, respectively. He was qualified ASRB-NET examination in 2016. Presently he is working as Assistant Professor at Department of Agricultural Chemistry and Soil science, College of Agriculture, Junagadh Agricultural University, Junagadh, Gujarat.

Mr. D. V. Hirpara: He was obtained B.Sc. (Hons.) Agri. and M.Sc. (Agri.) in Agril. Chem. & Soil Science degrees from Junagadh Agricultural University in 2015 and 2017, respectively. He was qualified ASRB-NET examination in 2018. Presently he is working as Senior Research Fellow at Department of Agricultural Chemistry and Soil science, College of Agriculture, Junagadh Agricultural University, Junagadh, Gujarat.

Dr. L. C. Vekaria: He has completed his Ph.D. in Agricultural chemistry and Soil Science from Junagadh Agricultural University, Junagadh (Gujarat). He is working as Assistant Research scientist in AICRP on Micronutrient project at Junagadh Agricultural University.

Dr. H. L. Sakarvadia: He has completed his Ph.D. in Agricultural chemistry and soil science from Junagadh Agricultural University, Junagadh (Gujarat). He has served as Assistant Professor for about 8 years in teaching, research, extension and technical works in Junagadh Agricultural University, Junagadh.

SOIL FERTILITY AND NUTRIENT MANAGEMENT

[A Way to Sustainable Agriculture]

A.S. Jadeja, D.V. Hirpara, L.C. Vekaria
and H.L. Sakarvadia

CRC Press is an imprint of the
Taylor & Francis Group, an **informa** business

NARENDRA PUBLISHING HOUSE
DELHI (INDIA)

First published 2021
by CRC Press
2 Park Square, Milton Park, Abingdon, Oxon, OX14 4RN
and by CRC Press
6000 Broken Sound Parkway NW, Suite 300, Boca Raton, FL 33487-2742

© 2021 Narendra Publishing House

CRC Press is an imprint of Informa UK Limited

The right of A.S. Jadeja et al. to be identified as authors of this work has been asserted by them in accordance with sections 77 and 78 of the Copyright, Designs and Patents Act 1988.

Reasonable efforts have been made to publish reliable data and information, but the author and publisher cannot assume responsibility for the validity of all materials or the consequences of their use. The authors and publishers have attempted to trace the copyright holders of all material reproduced in this publication and apologize to copyright holders if permission to publish in this form has not been obtained. If any copyright material has not been acknowledged please write and let us know so we may rectify in any future reprint.

All rights reserved. No part of this book may be reprinted or reproduced or utilised in any form or by any electronic, mechanical, or other means, now known or hereafter invented, including photocopying and recording, or in any information storage or retrieval system, without permission in writing from the publishers.

For permission to photocopy or use material electronically from this work, access www.copyright.com or contact the Copyright Clearance Center, Inc. (CCC), 222 Rosewood Drive, Danvers, MA 01923, 978-750-8400. For works that are not available on CCC please contact mpkbookspermissions@tandf.co.uk

Trademark notice: Product or corporate names may be trademarks or registered trademarks, and are used only for identification and explanation without intent to infringe.

Print edition not for sale in South Asia (India, Sri Lanka, Nepal, Bangladesh, Pakistan or Bhutan).

British Library Cataloguing-in-Publication Data
A catalogue record for this book is available from the British Library

Library of Congress Cataloging-in-Publication Data
A catalog record has been requested

ISBN: 978-1-032-06005-7 (hbk)
ISBN: 978-1-003-20023-9 (ebk)

Contents

Preface .... *vii*

1. Introduction 1
2. Plant Nutrients 16
3. Functions, Deficiency and Toxicity Symptoms of Plant Nutrients 36
4. Chemistry of Nutrient in Soils 50
5. Soil Fertility Evaluation 64
6. Soil Organic Matter (SOM) 76
7. Organic Manures 94
8. Chemical Fertilizers 133
9. Methods of Fertilizer Application 174
10. Problematic Soils 184
11. Fertilizer Storage and Fertilizer Control Order 229
12. Fertilizer Recommendations and Application 235
13. Assessment of Irrigation Water Quality 246
    References 259

Preface

The soil is the primary resource for life on this planet, but it is a very vulnerable resource. Increased food security and the alleviation of poverty require improvements in the productivity of food crops. The Indian population has been consistently increasing, and to fulfil their demand for food and raw materials, improved plant nutrition has remained as one of the major factors to increase crop yields. As a result, use of our knowledge of plant nutrition to maximize agricultural yields grows in importance. Smallholder farmers require simple and sustainable techniques to improve the productivity of crops. Fertilizer-use recommendations need to change with new developments, such as new varieties or better methods for assessing crop requirements. This information needs to reach the farmers and be implemented. Fertilizers have played an important role in increasing crop production. In India, the gap between potential and actual yield is very wide. Thus, developing effective and efficient soil fertility and plant nutrient management practices is indispensable for enhancing agricultural productivity and safeguarding the environment.

The present book is an excellent overview of Soil fertility and Nutrient management. The whole subject matters have been compiled in 13 chapters. Each chapter will emphasize on the mechanism of action and recent advances in the techniques for improvement of soil fertility and nutrient management. The outlooks of the authors are methodical and firm based on their own experiences during their carrier in the field of soil fertility management. With its application oriented and inter disciplinary approach, the book will be serve as useful reference to students, teachers, researchers scientists, policy makers and other interested in soil science, agronomy, crop science, environmental sciences and agriculture.

Authors

CHAPTER 1

INTRODUCTION

The relationship of soil, environment and society is intimate and depends on soil quality and its management. Soil is a naturally occurring thin layer of materials on the part of Earth's surface (land) composed of mineral and organic solids, gases, liquids and living organism which can serve as medium for plant growth. Soil science is the branch of agriculture that deals with soil considered as a natural body and as an important medium for plant growth. Therefore, a fundamental knowledge of soil science is prerequisite to meeting the natural resource challenges that will face humanity in the 21st century.

1.1 CONCEPT OF SOIL AND ITS DEFINITION

The term soil is derived from the Latin word "*Solum*" which means floor/ground. What a soil scientist calls soil - "a natural body on the earth's surface, a geologist may call fragmented rock, an engineer may call earth and economist may call land." There are two basic concepts of soil that have already evolved through two centuries of scientific study. The first one considered soil as a natural body, a biochemically weathered and synthesized product of nature and second one considers soil as a natural habitat for plants and other living organisms and justifies soil studies primarily on that basis.

The approaches of soil study: The two approaches: (i) pedological and (ii) edaphological approaches can be used in studying soils. In other words, there are two main branches of soil science are Pedology and Edaphology.

(i) *Pedological approach:* The origin of the soil, its classification, and its description are examined in pedology (from the Greek word "*pedon*", which means soil or earth). Pedology is the study of the soil as a natural body and does not focus primarily on the soil's immediate practical use. A pedologist studies, examines, and classifies soils as they occur in their natural environment.

2 SOIL FERTILITY AND NUTRIENT MANAGEMENT

(ii) Edaphological approach: Edaphology (from the Greek word *"Edaphos"*, which means soil or ground) is the study of soil from the standpoint of higher plants (Study of the soil in relation to plant growth, nutrition and yield of crops). Edaphologists consider the various properties of soils in relation to plant production. They are practical and have the production of food and fiber as their ultimate goal. To achieve that goal, Edaphologists must be a scientist to determine the reasons for variation in the productivity of soils and find means of conserving and improving productivity.

Functions of the soil: Followings are the major ecological functions of the soil

- Soils serve as medium for growth of all kinds of plants.
- Soils provide the physical support to the plants.
- Soils provide (supplies) moisture and nutrients for plant growth.
- It serves as a home for a myriad of organisms.
- Soils act as a living filter to clean water before it moves into an aquifer.
- It acts as a recycling system for nutrients and organic waste.
- It is a store house of nutrients.
- Soils modify the atmosphere by emitting and absorbing gases (carbon dioxide, methane, water vapor, and the like) and dust.

Soil is a three phase system: Soil mass is generally a three phase system as it consists of: (i) Solid phase, it includes various sizes of mineral and organic particles and living organisms (ii) Liquid phase, it includes water and dissolved nutrients/salts (soil solution) and (iii) Gases phase, it consisted of various gases like CO_2, N_2, O_2 *etc.* All these three phases play a very vital role for plant life. Relative proportion of these phases in soil governs the properties of soils. These properties of soil often determine the nature (type) of vegetation.

Soil as a medium for plant growth: In any ecosystem, soils play six key roles in relation to plant growth. These are as follows:

1. **Provide physical support:** Soils provide the physical support, anchoring the root systems so that the plant does not fall over or below away.
2. **Soil act as a ventilator:** To obtain energy, plant roots depend on the respiration process. In the respiration process, plant root consumes O_2 and release the CO_2. Soil pores allowing CO_2 to escape form root zones to atmosphere and fresh O_2 to enter the root zone.
3. **Provide water:** For growth, plant takes water from the soil through roots. Soil absorbs rain/irrigation water and holds it against the force of gravity in micro pores. Part of this water is used by plants.

4. **Act as temperature moderation:** The soil also moderates temperature fluctuations. The insulating properties of soil protect the deeper root system from extremes of hot and cold that often occur at the soil surface.
5. **Soil protects plant from toxins:** Several toxins (harmful substances) are produced in the soil as a result of microbial activity, root exudation, chemical reactions *etc.* or may result from human activity (pollution). Healthy soil will protect the plant from toxic concentration of such substances by ventilating gases, by decomposing or adsorbing organic toxins or by suppressing toxic producing organisms.
6. **Soil provides essential nutrients:** Seventeen elements have been shown to be essential elements (also called as essential nutrients), meaning that plants cannot grow and complete their life cycles without them. Soil provides most of essential nutrients to growing plants, without nutrients, plants life is not possible.

Soil properties and plant growth: Soil being a living organism exhibits all the three properties of any living being on the earth. These three properties are: physical properties, chemical properties and biological properties. All these three properties of the soils play an important role in determining its suitability for crop production. For example, water logged clay soil is suitable for rice cultivation whereas, remaining cultivated crops requires well drained soils. Citrus plants requires alight acidic to neutral soil reaction for optimum productivity. Moreover, one property of soil may also directly or indirectly affect other properties of soil. For example, neutral soil pH is ideal for soil fertility. Extreme acidic or alkaline pH adversely affects the soil fertility and ultimately crop productivity. Sandy soil has a poor CEC as well as fertility than the clay soils. Addition of Soil Organic Matter (SOM) improves the all physical, chemical and biological properties of soil.

1.2 SOIL FERTILITY AND PRODUCTIVITY

The interactions of various physical, chemical and biological properties in soil controls soil fertility (Plant nutrient availability). Therefore, understanding the soil properties and processes and how they are influenced by environmental conditions during growing season enables us to optimize nutrient availability and plant productivity.

Soil fertility: Soil fertility is the inherent capacity of soil to provide the essential plant nutrients in adequate amounts and in proper proportions for the plant growth. It represents the available nutrients status of the soil. The availability of nutrients to the plants is greatly influenced by physical (soil texture, soil structure, soil porosity *etc.*), chemical (pH, ESP, CEC *etc.*) and biological properties of soil.

Therefore, knowledge of soil fertility along with other properties of soil that affects the plant growth is essential to optimize the soil productivity.

Soil fertility and plant nutrition are two closely related subjects that emphasize the forms and availability of nutrients in soils, their movement in the soil and their uptake by roots. Soil fertility also can be readily altered by the application of soil amendments and nutrients *etc.* Knowledge of soil fertility is important for the development of soil management systems that produce profitable crop yields while maintaining soil sustainability and environmental quality.

Soil productivity: Soil productivity encompasses (involve) soil fertility plus all other factors affecting plant growth, including soil management practices. Therefore, Soil productivity is defined as the ability/capacity of the soil to produce crops yield with specific systems of management. OR soil productivity is the ability of the soil to produce the crop yield under specified management practices. It should be expressed in terms of yield (kg/ha. *etc.*).

Obtaining the maximum production potential of a particular crop in a region depends on the climatic condition, soil properties and biotic factors along with management practices. Thus, soil fertility is one of the several factors that affect soil productivity. All productive soils are fertile, but all fertile soils may or may not be productive due to some problems like water logging, alkalinity, salinity and adverse weather condition *etc.*

"All the productive soils are fertile but all the fertile soils may not be productive"

History of development of soil fertility

- **Francis Bacon (1591- 1624)** suggested that the principle nourishment of plants was water and the main purpose of the soil was to keep plants erect and to protect from heat and cold.
- **Jan Baptiste Van Helmont (1577 – 1644)** was reported that water was sole nutrient of plants.
- **Robert Boyle (1627 – 1691)** an England scientist confirmed the findings of Van Helmont and proved that plant synthesis salts, spirits and oil etc from H_2O.
- **Anthur Young (1741 – 1820)** an English agriculturist conducted pot experiment using Barley as a test crop under sand culture condition. He added charcoal, train oil, poultry dung, spirits of wine, oster shells and numerous other materials and he conducted that some of the materials were produced higher plant growth.

INTRODUCTION

- **Priestly (1800)** established the essentiality of O_2 for the plant growth.
- **J.B. Boussingault (1802-1882)** French chemist conducted field experiment and maintained balance sheet. He was first scientist to conduct field experiment. He is considered as father of field experiments.
- **Justus Von Liebig (1835)** suggested that......
    a. Most of the carbon in plants comes from the CO_2 of the atmosphere.
    b. Hydrogen and O_2 comes from H_2O.
    c. Alkaline metals are needed for neutralization of acids formed by plants as a result of their metabolic activities.
    d. Phosphorus is necessary for seed formation.
    e. Plant absorb every thing from the soil but excrete from their roots those materials that are not essential.

The field may contain some nutrient in excess, some in optimum and some in least, but the limiting factor for growth is the least available nutrient. The Law of Minimum, stated by Liebig in 1862, is a simple but logical guide for predicting crop response to fertilization. This law states that, "the level of plant production cannot be greater than that allowed by the most limiting of the essential plant growth factors". The contributions made by Liebig to the advancement of agriculture were monumental and he is recognized as the father of agricultural chemistry.

Crops depend on extrinsic and intrinsic factors for their growth and environment to provide them with basic necessities for photosynthesis. These essential plant growth factors include:

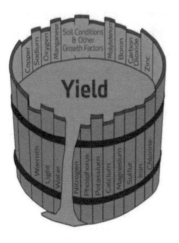

Justus Von Liebig's "Law of minimum published in 1873"
"if one growth factor/nutrient is deficient, plant growth is limited, even if all other vital factor/nutrients are adequate. Plant growth is improved by increasing the supply of the deficient factor/nutrient"

The Law of Minimum   Justus Von Liebig

Fig.1.1: The Law of Minimum

- light, heat, air, water, nutrients & physical support

  If any one factor, or combination of factors, is in limited supply, plant growth will be adversely affected. The importance of each of the plant growth factors and the proper combination of these factors for normal plant growth is best described by the principle of limiting factors. This principle states: "The level of crop production can be no greater than that allowed by the most limiting of the essential plant growth factors." The principle of limiting factors can be compared to that of a barrel having staves of different lengths with each stave representing a plant growth factor.

- **J.B. Lawes and J. H. Gilbert (1843)** established permanent manurial experiment at Rothemsted Agricultural experiment station at England. They conducted field experiments for twelve years and their findings were

  a. Crop requires both P and K, but the composition of the plant ash is no measure of the amounts of these constituents required by the plant.

  b. No legume crop require N. without this element, no growth will be obtained regardless of the quantities of P and K present. The amount of ammonium contributed by the atmosphere is insufficient for the needs of the crop.

  c. Soil fertility can be maintained for some years by chemical fertilizers.

  d. The beneficial effect of fallow lies in the increases in available N compounds in the soil.

- **S. N. Winogradsky** discovered the autotrophic mode of life among bacteria and established the microbiological transformation of nitrogen and sulphur. Isolated for the first time nitrifying bacteria and demonstrated role of these bacteria in nitrification (1890), further he demonstrated that free-living *Clostridium pasteuriamum* could fix atmospheric nitrogen (1893). Therefore, he is considered as "Father of soil microbiology".

- **Robert Warrington England** showed that the nitrification could be supported by carbon disulphide and chloroform and that it would be stopped by adding a small amount of unsterilized soil. He demonstrated that the reaction was two step phenomenons. First NH_3 being converted to nitrites and the nitrites to nitrates.

Soil fertility vs. Productivity: Basic difference between fertility and productivity of soil is given hereunder.

INTRODUCTION

Soil Productivity	Soil fertility
• It is the ability of soil to produce the crop yield (kg/ha) under specified management practices.	• It is the inherent properties of soil to supply essential plant nutrients in balanced proportion.
• Soil productivity is depends on many surrounding factors of the crop *i.e.* climatic, edaphic and biotic factors along with management practices followed by farmer.	• Soil fertility is only one of the several factors that affect soil productivity.
• Soil productivity = f (Climate, soil properties, biotic factors + management).	• Soil fertility = f (Available nutrient status of soil).
• All productive soils are fertile soil.	• All fertile soils may or may not be productive soil.
• It can be assessed in the field under climatic condition of a region.	• It can be analyzed in laboratory
• It is a broader term used to indicate crop yields	• It is considered as an index of available nutrient to plant

Factors affecting soil productivity

Obtaining the maximum yield of particular crop depends on growing season environment as well as soil properties and genetic potential of crop. In general, more than 50 factors affect crop growth and yield. Producer/consultant should have a skill to identify and minimize the factors that reduce the crop yield. Although, the producer cannot control many of the climatic factors, most of the soil and crop factors can and must be managed to maximize the productivity. For example, if the soil sodicity is the problem for crop production than producer must take appropriate measure to overcome the problems of soil sodicity for increasing the crop productivity. The important *factors that affect the soil productivity/ crop productivity* are given hereunder. (*Remember:* P = G x E x GE)

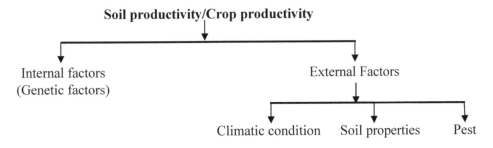

(A) Internal/biotic factors: Particular crop has their own genetic potential to grow well under given set of climatic condition. For example, apple favors temperate climate whereas, mango favors tropical climatic condition. Citrus crop requires well drained slight acidic to neural soils while, spota crop can be grown well under alkaline soil also. Some important internal as well as biotic factors are given below:

a. Crop species
b. Crop variety
c. Spacing
d. Geometry
e. Seed quality
f. Evapo-transpiration
g. Water availability
h. Nutrition

(B) External factors: The external factors are broadly grouped in to three categories *viz.* (i) Climatic factors, (ii) Edaphic factors and (iii) Pest, the details are as follow:

(i) Climatic factors: Climatic factors are those factors which are related to climatic conditions like rainfall, temperature *etc.* According to nature/adaptability of crop, all the below mentioned climatic factors should be optimal for obtaining maximum crop productivity.

a. Precipitation
  i. Quantity
  ii. Intensity
  iii. Distribution
b. Air temperature
c. Relative humidity
d. Light
  i. Quantity
  ii. Intensity
  iii. Duration
e. Wind
  i. Velocity
  ii. Distribution
f. CO_2 concentration

(ii) Edaphic OR Soil related factors: Following soil properties are greatly influence the soil productivity. Crop producer/consultant always gives more importance to soil properties for obtaining maximum crop productivity. Because it is possible to alter most of soil properties and make favorable them for better crop production. For example, soil salinity can be managed efficiently by providing the appropriate drainage technologies. Adverse soil pH (extreme acidic or alkali) may be altered by applying suitable soil amendments. Nutrient deficiency can also be corrected by supplying the nutrients from the outer sources *viz.* manures and fertilizers. Important soil properties that affect the soil productivity are listed below.

a. Soil Organic Matter (SOM)
b. Soil texture
c. Soil structure
d. Cation Exchange Capacity (CEC)
e. Base Saturation (BS)
f. Slope and Topography
g. Soil temperature
h. Soil aeration
  i. Soil depth (Root zone)
j. Soil organisms
k. Soil reaction
l. Soil related problems like salinity, alkalinity *etc.*
m. Soil management factors
  i. Tillage
  ii. Drainage
  iii. Others

(iii) Pest: Any organisms that damage the crop are termed as *pest*. For example, animals, humans, insect, weeds, diseases *etc.* that damage the crop. In modern agriculture, important weeds, insects, mites and diseases can be also managed efficiency with the use of agrochemicals that you will learn from relevant subject(s).

Factors affecting soil fertility

Soil/crop productivity is affected by surrounding environmental factors of crop *viz.* climatic parameters, soil properties and other biotic factors including management practices followed by producer *i.e.* weeding, irrigation, pest management *etc.*). It means soil fertility is only one of the several factors that affect soil productivity. Among the several soil properties, soil fertility is one property that represents the available nutrient status of soil during crop growth period. Moreover, the availability of nutrients in the soil and consequently their uptake by plant is influenced by other properties of soil as well as temperature and some biotic factors. The important properties that affect soil fertility are explained as below.

(A) Physical properties of soil
1. *Soil texture:* Soil texture is the relative percentages of soil separates *i.e.* sand, silt and clay) in the soil. The cations exchange capacity (CEC) of soil is positively associated with clay content thus; soil fertility. Therefore, sandy textured soils have low fertility than clay soils because sandy soils have a low CEC, WHC (water holding capacity) and SOC content than clay soils.
2. *Soil structure:* Soil structure is the arrangement of soil particles and their aggregates in defined pattern which facilities congenial environment for crop growth by providing balanced proportion of macro and micro porosity of soil. Therefore, well structured soil has a more fertile than poorly structured one because it facilitates both the water retention (micro pores) and soil aeration (macro pores) which are important for soil biochemical processes and nutrient availability which results in increases soil productivity.
3. *Soil water:* Optimum water content (60 to 70 % pores filled with water) in soil enhances the soil fertility. Soil water is a medium by with dissolved nutrients are absorbed by plant along with water.
4. *Soil porosity and aeration:* Balanced proportion of macro and micro pores are ideal for soil fertility and crop growth as it regulates both water retention and aeration status of soil.
5. *Other physical properties:* Other physical properties *viz.* water stable aggregates (WSA), infiltration rate, hydraulic conductivity, soil consistency; soil plasticity, soil compaction, soil crusting *etc.* also directly or indirectly affects the fertility as well as productivity of the soil.

(B) Soil reaction: Soil reaction (soil pH) is the single property which directly governs the nutrient availability and microbial processes in the soils. Slightly acidic to neutral soils (6.0 to 7.5) has good soil fertility. Acidic as well as alkaline reaction creates nutrient imbalances in the soil as strong acid soils may be deficient in N, P, K and Ca while, alkaline soils may be deficient in micronutrient cations and *vice-versa*.

(C) Soil Organic Matter (SOM): Addition of organic matter in the soil improves the fertility as well as productivity of the soil due to following reasons:
- Upon decomposition, it releases available form of plant nutrients in the soil.
- It also improves the physical properties of soil which indirectly increases the soil fertility.
- Organic matter also increases the microbial population and their activity in the soil thus increasing the soil fertility.
- During microbial decomposition of OM, some weak organic acids are released in the soil which solubilized the native unavailable nutrients into available form(s).

(D) Biological properties of soil: Relative activity of microorganisms plays a vital role in nutrient release (mineralization) in the soils. Especially N and S availability is highly governed by SOM and biochemical reactions in the soil. Nitrogen fixing bacteria *viz. Rhizobium spp., Azotobacter* fixes the atmospheric N in to the soil and increase the N availability. Phosphate solubilizing bacteria (PSB) increase the phosphorus availability in the soil whereas; Vesicular Arbuscular Mycorrhizae (VAM) fungi increase the P and Zn uptake by plants.

(E) Parent material and weathering processes: Natural supply of nutrients in the soil is closely tied up to parent material of that soil and vegetation under which it is developed. For example, apatite mineral is the main source of P in the soil while, K-feldspar and illite type silicate minerals are major sources of K in the soil. Whereas weathering processes is favored by humid and warm climate than the arid and cold region thus, the soils of arid and cold regions are inherently less fertile than humid warm region.

(F) Soil related problems: Some problems associated with soil viz. acidity, salinity, alkalinity, sodicity of soil decreased the fertility as well as productivity of soil. Therefore, understanding and management of such problems is required to increase fertility and productivity of soil.

(G) Other factors: Other factors *viz.* soil temperature, topography of soil, tillage operation, root growth and extension, cropping system and crop rotation also directly or indirectly influence the fertility and thus productivity of soil.

Management of soil fertility

There are following various ways by which producer can efficiently manage the soil fertility.

1. By maintaining the physical properties of soils.
2. Site specific nutrient management (applying nutrient based on the soil test).

3. Addition of nutrient through fertilizer in soil as well as foliar application as per crop needed/recommendation.
4. Addition of organic manures and green manuring in situ
5. Correction of soil related problems *i.e.* salinity, sodicity, acidity *etc.*
6. By preventing/minimizing the nutrient losses from the soils.

Features of a good soil management

Soil is defined differently for different purposes. It is the outer layer of earth crust capable of supporting plant growth. For an agronomist it is a natural body derived from rocks through weathering modified by physiography (relief) and organic matter over time. Weathering of rocks gives rise to soils. Depending on nature and properties of rocks and minerals and their response to weathering, the nature and properties of soils also vary. These properties are also modified by climate and biotic factors. Hence, soil is a function of climate, organic matter, relief, parent materials and time (s = f (cl. o.r.p.t). Parent material is in a way passive. Joffe stated "Different parent materials give rise to the same kind or type of soil whenever the principal factors of soil formation, the climate or biosphere, are the same". A corollary to this principle is that "similar parent materials give rise to a different kind or type of soil provided the principal factors are dissimilar". Depending on extent of weathering a soil is called a mature or immature soil. A mature soil is one that has reached the full development to be expected under existing biological and weathering processes. An immature soil is one that has not reached its final state of development. In a strict sense, no soil is mature because some change, development goes on. It is better to call them old or young soils. Oxisols are old soils and alluvial soils are young soils. Soil is foundation to every field/garden. Every healthy, productive field, yard and garden starts with healthy, productive soil. Preparing the soil properly makes more difference than any other thing you can do. You cannot put on enough nutrients and water to make up for poor soil. Since soils are so different in different areas, it is necessary to know what soil is, what your soil is like and what to add to improve it.

1.3 THE SOIL AS A NUTRIENT SOURCE FOR PLANTS

Mineral Nutrients in the Soil

- Both clay minerals and humic colloids have a negative net charge so that they attract and adsorb primarily cations. There are also some positively charged sites where anions can accumulate.

INTRODUCTION

- How tightly a cation is held depends on its charge and degree of hydration.
- In general, ions with high valences are attracted more strongly for example, Ca^{2+} is more strongly attracted than K^+. Among ions with the same valence those with little hydration are retained more firmly than those that are strongly hydrated.
- The tendency for cations adsorption decreases in the order Al^{3+}, Ca^{2+}, Mg^{2+}, NH_4^+, K^+ and Na^+

Adsorption and Exchange of ions in the soil: Both clay minerals and humic colloids have a negative net charge so that they attract and adsorb primarily cations. There are also some positively charged sites where anions can accumulate. How tightly a cation is held depends on its charge and degree of hydration. In general, ions with high valences are attracted more strongly for example, Ca^{2+} is more strongly attracted than K^+. Among ions with the same valence those with little hydration are retained more firmly than those that are strongly hydrated. The tendency for cations adsorption decreases in the order Al^{3+}, Ca^{2+}, Mg^{2+}, NH_4^+, K^+ and Na^+

Held more strongly						Held more weakly
$H^+ > Al^{3+} > Ca^{2+} > Mg^{2+} > NH_4^+ = K^+ > Na$

Fig.1.2: The tendency for cations adsorption (Lyotropic series soil).

The swarm of ions around particles of clay and humus as an intermediary between the solid soil phase and the soil solution.

- If ions are added to or withdrawn from the soil solution, exchange takes place between solid and liquid phases.
- Adsorptive binding of nutrient ions offers a number of advantages nutrients liberated by weathering and the decomposition of humus are captured and protected from leaching the concentration of the soil solution is kept low and

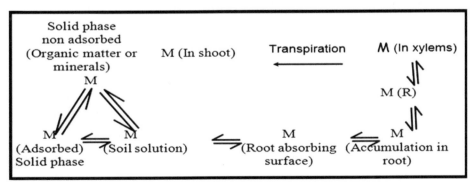

Fig.1.3: Nutrient release and path for absorption

14 SOIL FERTILITY AND NUTRIENT MANAGEMENT

relatively constant; so that the plant roots and soil organisms are not exposed to extreme osmotic conditions; when required by the plant, however, the adsorbed nutrients are readily available.

Plant obtains mineral nutrients through root uptake from the soil. Sources of these soluble nutrients in soil include:

1. ***Weathering of soil minerals:*** During weathering processes, nutrients are released in soil from parent materials.

2. ***Decomposition of soil organic matter (SOM):*** Plant residues, animal remains and microorganisms as well as addition of organic matter as manures are the principle sources of organic matter in the soil. During decomposition of OM, nutrients present in organic form are converted into inorganic form and becomes available to plants. For example, Protein N (organic-N) is converted in to NO_3^--N (Inorganic-N) through the series of biochemical reactions.

3. ***Fertilizer applications:*** Deficient nutrient can also be applied through inorganic fertilizers *viz.* urea, DAP, MOP *etc.*

4. ***Addition of organic manures:*** FYM, green manuring (GM) and many factory by-products are widely used as organic manures to improve fertility as well as productivity of the soil. Pressmud (PM), Biocompost (BC), Vermicompost (VC), Castor cake (CC) *etc.* are the major by-products of factory.

5. ***Atmospheric Nitrogen fixation by organisms:*** Some microorganisms have a capacity to fix atmospheric-N_2 into the soil and make it available to plants. For example, bacteria (*Rhizobium Azotobacter* and *Azospirillum*), actinomycetes (Frankia), cyanobacteria (Nostoc) and Azolla.

6. ***Inorganic industrial byproducts:*** Some industrial by-products like gypsum, distillery sludge, fly-ash *etc.* are widely used as a fertilizers or soil amendments.

7. ***Atmospheric deposition:*** Such as N and S from acid rain or N fixation by lightning discharge.

8. ***Sediment deposition:*** Deposition of nutrient rich sediments through erosion, flooding *etc.*

Losses of plant nutrients from the soil

Producer or agricultural consultant must know the below given several ways by which nutrients are lost from the field and thus, they should manage their field accordingly to reduce such losses of nutrients from the field.

1. ***Runoff losses:*** Loss of dissolved nutrients in water moving across the soil surface.

2. *Erosion losses:* Loss of nutrients in or attached to soil particles that are removed from the field by wind or water.
3. *Leaching losses:* Loss of dissolved nutrients with water that move down through the soil to ground water.
4. *Gaseous losses:* Primarily losses of different forms of N through volatilization as NH_3 (under alkaline soils) and denitrification as N_2O and N_2 (under water logged soils).
5. *Crop removal:* Plant uptake and removal of nutrients from the field in harvested products.
6. *Nutrient fixation:* In this process, nutrients are not lost from the soil but, sometimes readily available forms of nutrients are converted into unavailable form due to physico-chemical reactions.

CHAPTER 2

PLANT NUTRIENTS

As we know that all living organisms require food for survive, growth and reproduction. Therefore, every organism takes food and utilizes the food constituents for its requirements of growth and development. A series of processes are involved in the synthesis of food by plants, breaking down the food into simpler substances and utilization of these simpler substances for life processes. Nutrition in plants may thus be defined as a process of synthesis of food, its breakdown and utilization for various functions in the body.

Green plants synthesize their food materials from simple constituents (elements). Therefore, plants require a large number of elements which is taken up by the plants either from air and water or soil. In soil, the minerals (elements) are derived from minerals or are mineralized during the biological breakdown of organic matter. The mineral nutrients are taken up by the plants in the form of ions and incorporated into the plant structure or stored in the cell sap. One hundred and nine elements have been identified in the periodic table so far. Most of these exist in the earth's mantle, earth's crust and soil, though the magnitude of their occurrence varies. More than 60 elements are taken up by the plants. Based on their essentiality, they are grouped in to three categories **viz.** essential, beneficial and non essential. Before moving further, let us understand some of the key terms used in this chapter.

Element: It is the simplest form of the substance. It has definite properties and can not be change in other form. For example, carbon, nitrogen, phosphorus, potassium, iron, mercury *etc.*

Compound: When two or more elements combining to gather in a definite proportion and form third substance, the product is called compound. $C + O_2 = CO_2$.

PLANT NUTRIENTS

Nutrient: Nutrient may be defined as any mineral element/chemical compound required by an organism for their growth and development.

Nutrition: Nutrition is defined as the supply and absorption of chemical compounds required for plant growth and metabolism. OR It is a process of absorption and utilization of essential elements for plant growth and reproduction.

Essential nutrients: Essential nutrients are those nutrients/chemical compounds which are required by an organism for completion of their life cycle, without which organism is unable to complete its life cycle. They are also termed as essential elements.

Beneficial nutrients: Beneficial nutrients are the mineral nutrients which are stimulate plant growth, but are not essential or which are essential only for certain plant species, or under specific conditions. They are also termed as beneficial elements. For example, silicon (Si) is essential for paddy crop but it is not for all crop species.

Ballast element: - These elements most abundant in plant tissue and in earth crust also. It denoted as non essential element in plant e.g. Al (8 %), Si (27 %)

Non essential elements: Are those elements which are not required by plant for its normal growth.

2.1 ESSENTIAL NUTRIENTS

The essential nutrients may be defined as the nutrients which are required by plants for their normal growth and development. Without any of the essential nutrient, plant must be unable to complete its life cycle. For an element to be regarded as an essential nutrient, it must satisfy the following criteria, as given by Arnon and Stout (1939).

Criteria of nutrient essentiality

1. The plant must be unable to grow normally or complete its life cycle in the absence of an element (A deficiency of an element makes it impossible for plant to complete its life cycle).
2. The element is specific and cannot be replaced by another (The deficiency of the element is specific and can be prevented or corrected only by supplying that element).
3. The element is involved directly in the nutrition of the plant (The element plays a direct role in metabolism).

18 SOIL FERTILITY AND NUTRIENT MANAGEMENT

The seventeen nutrients recognized essential for plant growth are: carbon (C), hydrogen (H), oxygen (O), nitrogen (N), phosphorus (P), potassium (K), calcium (Ca), magnesium (Mg), sulphur (S), iron (Fe), Manganese (Mn), zinc (Zn), copper (Cu), nickel (Ni), boron (B), molybdenum (Mo) and chlorine (Cl). Nickel is lastly added in the list of essential nutrients. The requirement of molybdenum for plant is a very least as compared to all other essential nutrient. C, H, and O are not considered mineral nutrients but, most abundant elements in the plant as they constitute more than 90 per cent of plant biomass. Whereas, remaining all nutrients are considered as mineral elements/nutrients. Detailed information on these essential elements/nutrients pertaining to name of the discoverer(s), year of discovery, usable form and average concentration in plant are given in table 2.1.

Classification of essential nutrients: Essential nutrients/elements are classified based on: (i) sources (ii) requirement by the plant (iii) mobility in soil and in plant (iv) metal and non metal and (v) physiological functions and chemical behavior. The knowledge of these classifications is essential for efficient nutrients management.

(i) ***On the basis of their availability/source***: Based on their availability to crops, the essential nutrients are classified broadly in two groups:

Available from the air and water

Examples: C, H, O

Available from the s

Examples: N, P, K, S, Fe, Mn, Zn, Cu, Ni, Cl, E

Among the essential elements, non-mineral elements *i.e.* C, H and O are taken up by the plant either from air or water or both while remaining all essential nutrients are considered as mineral nutrients and they are available in the soil.

Table 2.1: Essential elements, their usable form, average concentration in plant tissue and essentiality discovered authors

Nutrients	Essentiality discovered authors (Discoverer)	Year of discovery	Plant Usable Form	Average conc. in plant tissue
C	Priestley *et. al.*	1800	CO_2	45.0%
H	Since time Immemorial	-	H_2O, H^+, OH^-	6.0%
O	Since time Immemorial	-	CO_2, O_2, H_2O	45.0%

[Table Contd.

Contd. Table]

Nutrients	Essentiality discovered authors (Discoverer)	Year of discovery	Plant Usable Form	Average conc. in plant tissue
N	Theodore de Saussure	1804	NO_3^-, NH_4^+	1.5%
P	C. Sprengel	1839	$H_2PO_4^-$, HPO_4^{-2}	0.2%
K	C. Sprengel	1839	K^+	1.0%
Ca	C. Sprengel	1839	Ca^{++}	0.5%
Mg	C. Sprengel	1839	Mg^{++}	0.2%
S	Sachs and Knop	1860	SO_4^{-2}	0.1%
Fe	E Gris	1843	Fe^{++}	100 ppm
Mn	J. S. McHargue	1922	Mn^{++}	20 ppm
Zn	A. L. Sommer and C. P. Lipman	1926	Zn^{++}	20 ppm
Cu	A. L. Sommer, C. P. Lipman and G. Mckinney	1931	Cu^{++}	6 ppm
Ni	P. H. Brown, R. M. Welch and E. E. Cary	1987	Ni^{++}	0.1 ppm
B	K. Warington	1923	H_3BO_3, $H_2BO_3^-$, HBO_3^{-2}, BO_3^{-3}	20 ppm
Mo	D. I. Arnon and P. R. Stout	1939	MoO_4^{-2}	0.1 ppm
Cl	T. C. Broyer *et. al.*	1954	Cl^-	100 ppm

(ii) *On the basis of nutrients requirement by plant*: Depending upon the quantity required by the plant, nutrients are classified into two broad groups: (i) macro and (ii) micro nutrients.

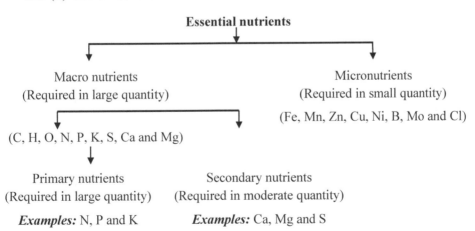

Macro nutrients: Macro nutrients or major nutrients are so called because these are required in large quantities, more than that of iron (Fe). Carbon (C), hydrogen (H) and oxygen (O) constitute 90 to 95 per cent of the plant dry matter weight and are mostly supplied through carbon dioxide (CO_2) and water (H_2O).

In addition, mineral nutrients *viz.* N, P, K, Ca, Mg and S which is taken up by the plant from soil are further sub divided in to two groups: (i) primary and (ii) secondary nutrients.

Primary nutrients: Nutrients which required in larger quantities are called primary mineral nutrients *viz.* N, P and K. The correction of their wide spread deficiencies is often necessary through application of commercial fertilizers of which these are the major constitutes.

Secondary nutrients: Nutrients which required in moderate quantities are called secondary mineral nutrients *viz.* Ca, Mg and S. The deficiency of secondary essential nutrients is not wide spread as like primary nutrients. But their localized deficiency often observed. For example, Ca deficiency is more common in strong acid soils. Their deficiency is corrected by applying deficient nutrient sources along with primary nutrients. For example, the phosphatic fertilizer, single super phosphate (SSP) contains both Ca and S. Like wise, ammonium sulphate, a nitrogenous fertilizer, also contains S.

Calcium (Ca) deficiency in acid soils: The deficiency of Ca is commonly observed under acidic soils especially in strong acid soils because of extensive leaching of salts as high rainfall. This deficiency is widely corrected by applying liming materials such as $CaCO_3$, $Ca(OH)_2$, and CaO, which not only increase the Ca availability but also increase the soil pH.

Primary nutrients	Secondary nutrients
• Mineral nutrients which required in larger quantity	• Mineral nutrients which required in moderate quantity
• Their deficiency is wide spread and N is most deficient nutrient across the world	• Localized deficiency observed
• Their deficiency is corrected by applying commercial fertilizes *viz.* Urea, DAP, MOP, SSP, Ammonium sulphate *etc.*	• Their source also available from fertilizers which are used for applying primary nutrients. For example: SSP, used as a P fertilizer, contains both Ca and S secondary nutrients.
• N, P and K are the primary nutrients	• Ca, Mg and S are the secondary nutrients

Sulphur (S) deficiency: In cultivated soils, the sulphur deficiency is ever increased day by day due to following reasons:

- An earlier, farmers were applied N, P and K nutrients through ammonium sulphate, single super phosphate (SSP) and potassium sulphate, respectively which also contains S. Whereas, recently farmers using Urea, Diammonium Phosphate (DAP) and Muriate of Potash (MOP) to supply N, P and K nutrients, respectively, which does not contain S.
- More nutrients are removed per year due to intensive agriculture and more so with cultivation of fertilizer responsive hybrid varieties.
- Addition of organic manures is not sufficiently available to correct nutrient deficiency.

Micronutrients: Nutrients that required in relatively smaller quantities but are essential as macronutrients are termed micronutrients. Some times they are also termed as minor or trace elements, but this label does not mean that they are less important than macronutrients. Micronutrient deficiency or toxicity can reduce plant yield just as macronutrient deficiency or toxicity does. Micronutrients are sub divided in to micro nutrient cations and micronutrient anions, depending upon the form in which plant absorb them.

Micronutrients

Micronutrient Cations
Examples: Fe, Mn, Zn, Cu and Ni

Micronutrient Anions
Examples: B, Mo and Cl

(iii) According to mobility: Essential nutrients and their compounds are further classified based on their mobility in the soil as well as within the different parts of plant.

(a) Nutrients mobile in soil: Based on relative ease of moment of nutrients within the soil or outside from the soil, they are classified in to three different groups: (i) high mobile, (ii) less mobile and (iii) Immobile.

1. **Mobile:** NO_3^-, SO_4^{2-}, BO_3^{3-}, Cl^- and Mn^{2-}
2. **Less mobile:** NH_4^+, K^+, Ca^{2+}, Mg^{2+} and Cu^{2+}
3. **Immobile:** $H_2PO_4^-$, HPO_4^{2-} and Zn^{2+}

This classification is very useful for deciding the methods of fertilizers application For example, N is mobile nutrient in the soil and thus applied periodically at critical growth stages while; P is immobile in the soil, so all recommended dose of P should be applied once at the time of sowing or planting.

(b) Nutrients mobile in plant: According to movement of nutrients within the plant, they are divided into four groups which are as follows:
1. **Highly mobile:** N, P and K
2. **Moderately mobile:** Zn
3. **Less mobile:** S, Fe, Mn, Cu, Mo and Cl
4. **Immobile:** Ca and B

This classification is also useful for deciding the methods of fertilizer application as well as for diagnosing the deficiency symptom(s) of nutrients in the plant. For example, foliar application of boron (B) is advised in case of fruit cracking for quick result because boron (B) is immobile in plant. Moreover, the deficiency symptoms of N, P and K (mobile nutrients) are first observed at older leaves of plant because they are easily translocated from older leaves to younger leaves as they are mobile in the plant. In contrast, deficiency of Ca and B are first observed at growing tips because of they are immobile in the plant thus not easily translocated from older leaves to growing tips.

(iv) Based on metal and non metal nature of nutrients
1. **Metal:** K, Ca, Mg, Fe, Mn, Zn and Cu
2. **Non metal:** N, P, S, B, Mo and Cl

(v) According to Cation and Anion
1. **Cation:** K^+, Ca^{++}, Mg^{++}, Fe^{++}, Mn^{++}, Zn^{++}, Cu^{++} *etc.*
2. **Anion:** NO_3^-, $H_2PO_4^-$, SO_4^{-2} *etc.*

(vi) According to biochemical behaviour and physiological function

Mengal and Kirby classified plant nutrients on the basis of their bio-chemical behavior and physiological function.

From a physiological point of view, it is difficult to justify the classification of plant nutrients in to macronutrients and micronutrients. Therefore classification of plant nutrients according to biochemical behaviour and physiological function seems more appropriate. So the classification of plant nutrients is shown below.

Essential plantnutrients	Biochemical functions
Ist Group C, H, O, N, S	• Major constituents of organic material
	• Involved in enzymatic processes
	• Assimilation by oxidation - reduction reaction
IInd Group P, B, Si	• Esterification with native alcohol groups in plants

Essential plantnutrients	Biochemical functions
	• Involved in energy transfer reactions (phosphate esters)
IIIrd Group K, Na, Mg, Ca, Mn, Cl	• Non specific functions establishing osmotic potentials.
	• Enzyme activation
	• Bridging of reaction partners
	• Balancing ions in plants (Maintain the reaction)· Control membrane permeability and electro – potentials
IVth Group Fe, Cu, Zn, Mo	• Present in chelated form in plant and incorporated in prosthetic groups
	• Enable electron transport by valence change

Terminology related to mineral nutrients: Following terms are routinely used in soil fertility and nutrient management field.

Available nutrient: The portion of any essential nutrient or compound in the soil which can be easily absorbed (taken up) by the plant is termed as available nutrient. In soil, total amount of any nutrient present in different forms. All these forms may not be absorbed by plant roots as plant absorbs only definite form(s) of nutrient (see table 1). Therefore, only fraction/form(s) of nutrient which is easily absorbed by plant root is termed as available nutrients and they are designated as available N, available P_2O_5 *etc.* For example, total amount of N present in soil in different forms *viz.* Organic N (protein, amino acids, amine *etc.*) and inorganic N (mostly as NH_4^+, and NO_3^-). Among which, only inorganic fractions of N is termed as available N and it is expressed in terms of kg ha^{-1}, in case of major nutrients (large quantity) or mg kg^{-1} soil, in case of micronutrients (small quantity).

Trace elements: The elements which are found in very low concentration, perhaps less than one ppm or still less in soil, plant, water and other materials is termed as trace elements.

Heavy metal: A metal having specific gravity more than 5.0 or having atomic number higher than 20 is termed as heavy metal. As a corollary, any metal heavier than calcium (Ca) is a heavy metal. For example: Pb, Cr, Co *etc.*

Nutrient content: Concentration of nutrient or its amount per unit weight of a plant is termed as nutrient content. It is expressed in terms of percentage (in case of large amount) or ppm (in case of small amount). For example: Mango leaves contain 1.0 to 1.5 % N and 20 to 50 ppm Fe.

Nutrient accumulation: Storage of nutrient in a particular part or portion of the plant is called nutrient accumulation. For example: Cowpea pods contain 5.0 % N.

Nutrient uptake: Amount of nutrient taken up by the growing crops from either soil or other sources, is called nutrient uptake. It is expressed in terms of kg ha^{-1} or g ha^{-1}. For example: N uptake by brinjal crop is 50 kg ha^{-1}.

$$\text{Nutrient uptake (kg ha}^{-1}\text{)} = \frac{\text{Nutrient content (\%)} \times \text{Yield (kg ha}^{-1}\text{)}}{100}$$

Nutrient Removal: The nutrient contained in the harvest portion of the crop is termed as nutrient removal.

Nutrient Fixation: Fixation is the process whereby readily available form(s) of plant nutrients are converted in to unavailable form **OR** Fixation is the process whereby readily available soluble plant nutrients are converted in to less soluble forms by reaction with inorganic or organic compounds of the soil restricting their mobility in the soil and thereby suffer a decrease in their availability to the plant.

Terms used to indicate the concentrations of nutrient (nutrient content) in plant

Sufficient: The range of nutrient content in plants associated with optimum crop yields called sufficient nutrient.

Insufficient: Nutrient contents associated with only growth reductions and not accompanied by appearance of deficiency symptoms are termed as insufficient element **OR** when the concentration of an essential element is below than that required for optimum yield or when there is an imbalance with another element.

Deficient: When an essential nutrient is at a very low concentration that severely reduced the crop yield and produce more or less distinct deficiency symptoms is called deficient nutrient.

Excessive: When a concentration of an essential nutrient is sufficiently high to result in a corresponding shortage of another nutrient.

Toxic: When an essential nutrient at a to high concentration that severely limits yield and produce more or less distinct toxicity symptoms is called toxic nutrient. Severe toxicity will result in plant death.

PLANT NUTRIENTS

Now, we understand above terms properly with following example:

Suppose, **1.0 to 1.3 % N** in mature tomato leaves is required for optimum tomato yield then,

Sr. No.	N content in tomato leaves	Term used	Remarks
1.	1.0 to 1.3 % N	Sufficient	Adequate concentration of N to produce optimum tomato yield.
2.	0.8 % N	Insufficient	Inadequate (low) concentration of N than required to produce optimum tomato yield. At this concentration, yield of tomato is reduced but, plant does not show any deficiency symptoms. This condition also called as hidden hunger.
3.	> 0.5 % N	Deficient	Very low concentration of N than required to produce optimum tomato yield. At this concentration, yield of tomato is greatly reduced and plant shows deficiency symptoms.
4.	1.5 % N	Excessive	High concentration of N than required concentration to produce optimum tomato yield. At this concentration, yield of tomato is reduced due to imbalances of nutrients but, plant does not show any toxicity symptoms.
5.	< 2.0 % N	Toxic	Very high concentration of N than required concentration to produce optimum tomato yield. At this concentration, yield of tomato is greatly reduced due to imbalances of nutrients and plant shows toxicity symptoms.

Relationship between essential nutrient concentration and plant growth

Optimum and balanced proportion of nutrients are required for optimum plant growth. As the nutrient concentration increases to ward the critical level, plant yield increases. Above the critical level the plant contains sufficient levels for normal growth and can continue to absorb nutrients without increasing yield (luxury consumption). Excessive absorption of a nutrient or element can be toxic to plant and reduce yield.

Figure 2.1. shows that yield is severely affected when a nutrient is deficient, and when the nutrient deficiency is corrected, growth increase more rapidly than nutrient concentration. Under severe deficiency, rapid increase in yield with added nutrient can cause a small decrease in nutrient concentration is called Steenberg effect or Dilution effect. The Steenberg effect results from dilution of the nutrient in the plant by rapid plant growth. When the concentration reaches the critical range, plant yield is generally maximized. Nutrient sufficiency occurs over a wide concentration range, where yield is unaffected. Increase in nutrient concentration above the critical range indicate that the plant absorbing nutrients above that needed for maximum yield is called luxurious consumption. Elements absorbed in excessive quantities can reduce plant yield directly through toxicity or indirectly by reducing concentration of other nutrients below their critical ranges.

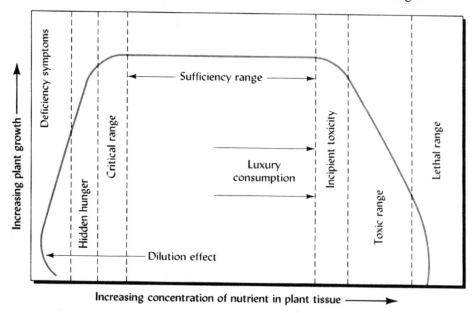

Fig. 2.1: Relation between essential plant nutrient concentration and plant growth or yield

2.2. BASIC SOIL -PLANT RELATIONSHIPS

Mineral Nutrients in the Soil: Mineral nutrients occur in the soil in both dissolved and bound form. Only a small fraction (less than 0.2%) of the mineral nutrient supply is dissolved in soil water. Most of the remaining *i.e.,* almost 98 % is either bound in organic form, humus and relatively insoluble inorganic compounds or incorporated in minerals. These constitute a nutrient reserve, which becomes available very slowly as a result of weathering and mineralization of humus. The remaining 2.0 % is adsorbed on soil colloids. The soil solution, the soil colloids and the reserves of mineral substances in the soil are in a state of dynamic equilibrium, which ensures continued replenishment of supplies of nutrient elements.

Nutrient dynamics in the soil: Plant absorbs dissolved nutrients along with water from the soil solution through roots. Therefore, sufficient concentration of essential nutrients must be maintained in soil solution. But, there are different sites of nutrient reserve in soil *viz.* soil solution, exchange complex, minerals and organic matter *etc.* The ionic equilibration is maintained among them especially between soil solution and exchange complex. Thus, it is essential to know nutrient dynamics, the relationship between these nutrient reservoirs and important physical, chemical and biological processes that releases the nutrients in soil solution and consequently uptake by plants.

The interaction of various physical, chemical and biological properties in soils controls plant nutrient availability. Understanding these processes and how they are influenced by environmental conditions during the growing season enables us to optimize nutrient availability and plant productivity.

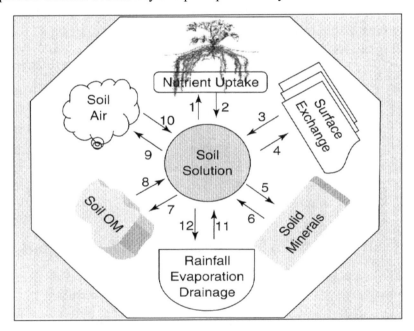

Fig. 2.2: Various dynamic soil processes influencing the nutrient availability

Nutrient supply to plant roots is a very dynamic process (Fig. 2.2). Plant nutrients are *absorbed* from the soil solution and *release* small quantities of ions (H^+, OH^- and HCO_3) back to the solution by plant roots (reaction 1 and 2). As a result several chemical and biological reactions occur to buffer or resupply the solution. Ions *adsorbed* to the surface of minerals *desorbs* from these surfaces to resupply the solution (reaction 3 and 4). Ion exchange (*adsorption and desorption*) in soil is an important reaction to plant nutrient availability. Soils also contain minerals that can *dissolve* to resupply the soil solution (reaction 6). Addition of the nutrients through fertilization or other inputs increases the nutrient concentration in the soil solution. Although some of added ions remain in solution, some are *adsorbed* to the mineral surfaces (reaction 4) or *precipitated* as solid minerals (reaction 5).

Microbial reactions are important to plant nutrient availability as well as other properties related to soil productivity. As soil organisms degrade plant residue (organic matter) they can *absorb* ions from the soil solution in to their tissues (reaction 7). When organisms die, they *release* nutrients back to the soil solution (reaction 8). Plant roots and soil organisms utilize O_2 and respire CO_2 through metabolic activity (reaction 9 and 10). As a result, CO_2 concentration in the soil air is greater than in atmosphere. Diffusion of gases in soil decreases dramatically with increasing soil water content.

Numerous environmental factors and human activities can influence ion concentration in the soil solution (reaction 11 and 12). For example, adding P fertilizer to the soil initially increases the $P_2O_4^-$ concentration in the soil solution. With time, the $P_2O_4^-$ concentration will decreases with plant uptake (reaction 1), $P_2O_4^-$ adsorption on mineral surfaces (reaction 4) and P mineral precipitation (reaction 5).

All these processes and reactions are important to plant nutrient availability; however, depending on the specific nutrient, some processes are more important than others. For example, microbial processes are more important to *N and S* availability than then mineral surface exchange reactions, whereas the opposite is true for *K, Ca and Mg*.

Important terminology of nutrient dynamics in soil

Exchange complex/site: It is the chemically active surface area of soil colloids as it possess charge (-and +) where exchange of ions takes place.

Surface exchange/Ion exchange: Exchange of ions between solid and liquid phase of soil. If the exchanged ions are cations then called as cations exchange and if anions then called as anion exchange.

Cation exchange capacity (CEC): It may be defined as the amount of exchangeable cations per unit weight of soil. It is expressed as me/100 g soil or $cmol_e$ (P^+)/ kg soil (SI unit).

Anion exchange capacity (AEC): It may be defined as the amount of exchangeable anions per unit weight of soil. It is also expressed as me/100 g soil or $cmol_e$ (P^+)/ kg soil (SI unit).

Adsorption: Adsorption is a phenomenon of surface adherence due to electrostatic bond between charge on soil colloids and ions. For example: Ca ion adsorbed on soil colloids.

Desorption: Release of adsorbed ion from exchange complex to soil solution

Precipitation: It is process in which two or more elements/compounds react with each other and form third insoluble compound(s). It is also called as chemical nutrient fixation. In this process, the readily available form of nutrients is converted in to insoluble form thus fertility of soil is decreased.

For examples: $Ca + PO_4 = (Ca)_3(PO_4)_2$, $Fe + PO_4 = FePO_4$.

Mineralization: It is the process in which organic form of nutrients are converted in to inorganic form through series of biochemical reactions. For example: Protein-N (organic) = NH_4^+-N (inorganic)

Immobilization: It is the process in which inorganic form of nutrients are converted in to organic form through the organisms. For example: NH_4^+-N/NO_3^--N (inorganic) = Protein-N (organic).

The mechanism of nutrient movement

The mechanism of nutrient movement from soil to plant body has been given by Fried and Broeshast (1967). This mechanism involves very complex processes/phenomenon in soil and plant body. Here, the simplify system explains that how the nutrients are move between the phases of soil and from soil solution to plant roots and then shoots via xylem through transpiration pull of water (Fig. 2.3).

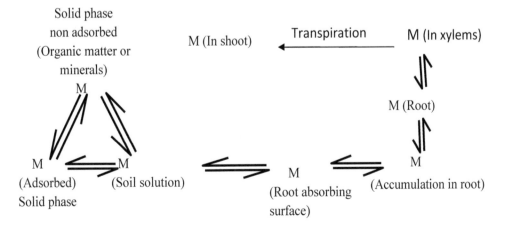

Fig. 2.3: Nutrient release in soil and path for absorption in plant.

2.3. MECHANISMS OF NUTRIENT TRANSPORT FROM SOIL TO PLANT ROOT

Prior to absorption of different plant nutrients, they reach the surface of roots by the following two theories/mechanisms.

(a) Soil Solution theories: Following two important processes of soil solution theory

(i) **Mass flow:** Movement of ions and other dissolved substances along with moving water to the root zone is called mass flow. (Due to transpirational water uptake by the plant), *e.g.:* Nitrogen

Factors affecting mass flow

a. Soil water content
   i. Dry soil where there is no nutrient movement
b. Temperature
   i. Low temperature reduces transpiration and evaporation
c. Size of root system
   i. Affects water uptake and therefore movement
   ii. Root density much less critical for nutrient supply by mass flow than for root interception and diffusion

(ii) **Diffusion:** Diffusion occurs when there is concentration gradient of nutrients between root surface and surrounding soil solution. Ions move from the region of higher concentrations to the region of low concentration is called diffusion. *e.g.:* Phosphorus, Potassium

Factors affecting diffusion:

A. **Fick's law is given as**

$$dC/dt = De * A * dC/dX$$

dC/dt = diffusion rate (change in concentration over time)

De = effective diffusion coefficient

A = cross sectional area for diffusion

dC/dX = concentration gradient (change in concentration over distance)

Diffusion rate is directly proportional to concentration gradient, diffusion coefficient, and the area available for diffusion to occur

B. *Effective diffusion coefficient:*

Effective diffusion coefficient

$$De = Dw * q * (1/T) * (1/b)$$

PLANT NUTRIENTS 31

Where,

D*w* = diffusion coefficient in water

q = volumetric soil water content

T = tortuosity factor

b = soil buffering capacity

a. *Diffusion coefficient in water (Dw)*
   i. Includes a temperature factor
   ii. Colder = slower diffusion
b. *Soil water content*
   i. Drier soil = slower diffusion
   ii. Less water = less area to diffuse through
c. *Tortuosity*
   i. Pathways through soil are not direct
   ii. Around soil particles, through thin water films
   iii. Affected by texture and water content
       1. More clay = longer diffusion pathway
       2. Thinner water films = longer path
d. *Buffering capacity*
   i. Nutrients can be removed by adsorption as they move through soil, reducing diffusion rate

C. **How far can nutrients diffuse in a growing season?**
   a. *Diffusion distances are very short*
      i. K ~ 0.2 cm
      ii. P ~ 0.02 cm
   b. *Size and density of plant root systems is very important for nutrients supplied by* diffusion
   c. *Has implications for fertilizer placement*

(b) Contact exchange theory (Root interception): A close contact between root surface and soil colloids allows a direct exchange of ions is called root interception. Contact exchange theory has a less importance than soil solution theory (mass flow and diffusion) and it is not well understood.

Factors affecting root interception
a. Anything that restricts root growth
   i. Dry soil
   ii. Compaction
   iii. Low soil pH
   iv. Poor aeration
   v. Root disease, insects, nematodes
   vi. High or low soil temperature
b. Root growth is necessary for all three mechanisms of nutrient supply, but absolutely essential for root interception to occur

2.4. SOIL REACTION AND BUFFERING:

- Soil pH or soil reaction is an indication of the acidity or alkalinity of soil and is measured in pH units. The pH scale goes from 0 to 14 with pH 7 as the neutral point. As the amount of hydrogen ions in the soil increases, the soil pH decreases, thus becoming more acidic. From pH 7 to 0, the soil is increasingly more acidic, and from pH 7 to 14, the soil is increasingly more alkaline or basic.

- Sorenson (1909) defined the pH and gives the pH scale. Using a strict chemical definition, pH is the negative log of hydrogen ion (H^+) activity in an aqueous solution in moles/L.

- The point to remember from the chemical definition is that pH values are reported on a negative log scale. So, a 1 unit change in the pH value signifies a 10-fold change in the actual activity of H^+, and the activity increases as the pH value decreases. To put this into perspective, a soil pH of 6 has 10 times more hydrogen ions than a soil with a pH of 7, and a soil with a pH of 5 has 100 times more hydrogen ions than a soil with a pH of 7. Activity increases as the pH value decreases.

$$pH = -\log_{10}(H^+)$$

Where: (H^+) is the activity of hydrogen ions in moles/lit.

Pure water is weakly dissociated in to H^+ and OH^- ions according to following equations

$$H_2O = H^+ + OH^-$$

According to law of dissociation

$$[H^+] \times [OH^-] / H_2O = K$$

Where;

H^+ etc are the concentration and K is the dissociation constant. Since concentration of undissociated water remains practically the same because of very little ionization of H_2O molecules the above relationship becomes:

$[H^+] \times [OH^-] = Kw = 10^{-14}$ at 200 ^{0}C Kw=Ion product constant of water

At neutrality $H^+ = OH^-$ and $H^+ = 10^{-7}$ or pH=7. Pure water has a pH value of 7. As the hydrogen activity increases the pH value will decrease while it will go up with rise in hydroxyl ion activity.

Importance of Soil pH

- Suitability of soil for crop production
- Availability of soil nutrients to plants
- Microbial activity in the soil
- Lime and gypsum requirement of soil
- Physical properties of soil like structure, permeability etc.

EFFECT OF SOIL pH ON NUTRIENT AVAILABILITY

The availability of plant nutrients are more at a pH range of 6-7 except Mo

Nitrogen

One of the key soil nutrients is nitrogen (N). Plants can take up N in the ammonium (NH_4^+) or nitrate (NO_3^-) form.

- At pH near neutral (pH 7), the microbial conversion of NH_4^+ to nitrate (nitrification) is rapid, and crops generally take up nitrate. In acid soils (pH < 6), nitrification is slow, and plants with the ability to take up NH_4^+ may have an advantage

Phosphorus

The form and availability of soil phosphorus is highly pH dependent.

- When the soil is neutral to slightly alkaline, the HPO_4^- ion is the most common form. As the pH is lowered both the HPO_4^- and H_2PO4^- ion prevail. At higher acidities H_2PO4^- ions tends to dominate. The most plants absorb phosphorus in HPO_4^-.
- Between pH 6-7, phosphorus fixation is at minimum and availability to higher plants is maximum.

Potassium

The fixation of potassium (K) and entrapment at specific sites between clay layers tends to be lower under acid conditions. This situation is thought to be due to the presence of soluble aluminium that occupies the binding sites.

Calcium, Magnesium and Sulphur

The availability of Ca and Mg is more above pH 7.0.

Sulphate (SO_4^{2-}) sulphur, the plant available form of S, is little affected by soil pH.

Micronutrients

The availability of the micronutrients manganese (Mn), iron (Fe), copper (Cu), zinc (Zn), and boron (B) tend to decrease as soil pH increases.

- The exact mechanisms responsible for reducing availability differ for each nutrient, but can include formation of low solubility compounds, greater retention by soil colloids (clays and organic matter) and conversion of soluble forms to ions that plants cannot absorb.
- Molybdenum (Mo) behaves counter to the trend described above. Plant availability is lower under acid conditions.

Soil pH and soil organisms

- Growth of many bacteria and actinomycetes is inhibited as soil pH drops below 6
- Fungi grow well across a wide range of soil pH
- Therefore fungi are dominant under acid conditions
- Less competition from bacteria and actinomycetes
- Earthworms do best when soil pH >6.5
- Nitrification greatly inhibited at pH <5.5
- N fixation greatly restricted a pH <6
- Decomposition of plant residues and OM may be slow in acid conditions (pH <5.5)

Soil Buffering Capacity

The ability to resist a change in pH refers to buffering capacity of the soil

- The buffering capacity increases as the cation exchange capacity increases. Thus, heavier the texture and the greater the organic matter content of a soil, the greater is the amount of acid or alkaline material required to change its pH.
- The colloidal complex acts as a powerful buffer in the soil and does not allow rapid and sudden changes in soil reaction.
- Buffering depends upon the amount of colloidal material present in soil. Clay soils rich in organic matter are more highly buffered than sandy soils.
- Importance of buffering in agriculture
- The stabilization of soil pH through buffering act as a effective guard against deficiency of certain plant nutrients and excess availability of others in toxic amounts which would seriously upset the nutritional balance in the soil.

FACTORS INFLUENCING NUTRIENT AVAILABILITY:

Several factors influence nutrient availability:

1. Natural supply of nutrients in the soil which is closely tied up to parent material of that soil and vegetation under which it is developed.
2. Soil pH also affects nutrient release,
3. Relative activity of microorganisms which play a vital role in nutrient release and may as in the case of Mycorrhizia directly function in nutrient uptake
4. Fertility addition in the form of commercial fertilizer, animal manure and green manure, etc
5. Soil temperature, moisture and aeration.

CHAPTER 3

FUNCTIONS, DEFICIENCY AND TOXICITY SYMPTOMS OF PLANT NUTRIENTS

3.1. IMPORTANT FUNCTIONS AS WELL AS DEFICIENCY AND TOXICITY SYMPTOMS OF ESSENTIAL MINERAL NUTRIENTS

NITROGEN: Scarcity of nitrogen (N) is the most widely occurring nutritional limit on the productivity of crops across the word. Nitrogen is also the element most widely over applied to agricultural crops and the most widely responsible for water quality. Nitrogen is absorbed by plant roots as NO_3^- and in case of rice, as NH_4^+ also. In the N sufficient plants, its concentration varies from 1.0 to 6.0 per cent. This nutrient is most often deficient in the soil and is the one which frequently creates serious nutritional problems.

Functions of nitrogen in plant: Following are the major functions of N in relation to plant growth.

- Nitrogen is a basic constituent (part) of "life". Because it is an essential component of amino acids, proteins, nucleic acids, porphyrins, nucleotides, alkaloids, enzymes, hormones, vitamins *etc.*
- Nitrogen is also an integral part of chlorophyll, which is a primary absorber of light energy needed for photosynthesis.

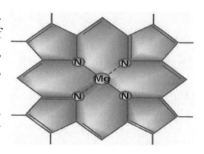

(The basic unit of chlorophyll's structure is the porphyrins ring system, composed of four pyrrole rings, which contains one nitrogen atom and four carbon atoms. A single magnesium atom is bounded in the centre of each porphyrins ring.)

- Nitrogen is responsible for the transfer of genetic code to the off-springs. Because of it is a constituent of nucleic acids *viz.* DNA and RNA.
- Nitrogen also imparts vigorous vegetative growth and dark green colour to plants.
- Nitrogen improves the quality of leafy vegetables and fodders.
- It produces early growth and also results delay in maturity.
- It improves the protein quality of the plant.

The supply of nitrogen is related to carbohydrate utilization. When nitrogen supplies are insufficient, carbohydrates will be deposited in the vegetative cells, which will cause them to thicken. When nitrogen supply is optimum and conditions are favourable for growth, proteins are formed from the manufactured carbohydrates. Less carbohydrate is thus deposited in the vegetative portion, more protoplasm is formed, and, because protoplasm is highly hydrated, more succulent plant results. Excessive supply of nitrogen develops excessive succulence which results harmful effects in some crops like weakening of fiber in cotton, lodging in case of grain crops *etc.*

Deficiency Symptoms of nitrogen: Under N deficit soil, plant shows following deficiency symptoms:

- Stunted plant growth observed due to N deficiency.
- Older leaves become light green to yellow colour (chlorosis in older leaves). Because N is highly mobile in plant thus it easily translocated from younger to older leaves.

Chlorosis: Chlorosis is the yellowing of plant tissue due to limitations on chlorophyll synthesis. This yellowing can be generalized over the entire plant, localized over entire leaves or isolated between some leaf veins (i.e. interveinal Chlorosis).

- Reduction in flowering and yields.
- Due to N deficiency, protein content in plant decreased. Because N is the main component of protein. It is assumed that protein contains 16.0 % N.

Therefore, Protein content (%) = N content (%) × 6.25

16 g N = 100 g protein

1.0 g N = 100 x 1.0/16 = 6.25

Toxicity/Excessive consumption of nitrogen:
- Excess application of N caused the succulent plant growth therefore lodging occurs.
- It produces early and vigorous growth
- It delays reproductive phase and maturity of plants
- Due to succulence of plants and dark green vigorous growth, plant becomes more susceptible to insect, pest and disease attacks.

PHOSPHORUS: Phosphorus (P) is the second most widely limiting nutrient after nitrogen. In modern agriculture, over application of P fertilizers are responsible for widespread pollution of aquatic systems, *i.e. Eutrophication*. Plant roots mostly absorb phosphorus as the dihydrogen orthophosphate ($H_2PO_4^-$) ion, but under neutral to alkaline conditions, it is also taken up as monohydrogen orthophosphate (HPO_4^{-2}) ion. In normal P-sufficient plants, P content varies from 0.1 to 0.4 per cent by weight.

Functions of phosphorus in plant:
- Phosphorus has a great role in energy storage and transfer.
- Phosphorus is a constituent of nucleic acid, phytin, and phosphor-lipids.
- It is also an essential constituent of majority of enzymes which are great importance in the transformation of energy, in carbohydrate metabolism and in respiration of plants.
- It is closely related to cell division and development.
- Phosphate compound (ADP and ATP) act as "energy currency" within the plants.
- It stimulates early root development and growth and thereby helps to establish seedlings quickly.
- It increases straw strength in cereals.
- It brings early maturity of crops and counteracts the effect of excessive nitrogen.
- It increases the N_2 - fixation capacity of legume crops.

FUNCTIONS, DEFICIENCY AND TOXICITY SYMPTOMS OF PLANT NUTRIENTS

Deficiency symptoms of phosphorus:

- Because of its faster mobility in the plants, P gets easily translocated from older tissue to the meristematic tissues. Therefore, deficiency symptoms of P appear first on the older leaves.
- Phosphorus deficiency overall associated with the overall stunting plant growth and a darker green colouration of leaves.
- Under sever deficiency, older leaves become bronzed or develop reddish - purple tips and leaf margins.
  (The purple colour is due to accumulation of sugars that enhances synthesis of anthocyanin, a purple pigment) in the leaf.)
- Due to P deficiency, severe restriction occurs in the growth of plant tops and roots.
- In general, P deficient plants are thin, erect and spindly with sparse and restricted foliage.

Toxicity of phosphorus: There is no visual toxicity symptoms of P are observed in plant, however excess p availability in soil of P fertilizer application may cause micronutrient deficiency, especially iron or zinc due to fixation of Fe and Zn by phosphate compounds.

POTASSIUM: Potassium (K) is absorbed by the plants in larger amounts than any other nutrient except N. Plant root absorbs potassium as potassium ion (K^+). Its concentration in healthy plant tissues varies from 1.0 to 5.0 per cent. Potassium is a unique element in the sense that plant can accumulate it in abundant amounts without exhibiting any toxicity symptoms. Unlike N, P and most other nutrients, K is not structural component of the plant.

Functions of potassium in plant

- Opening and closing of stomata: It regulates the opening and closing of the stomata, which are essential for photosynthesis, water and nutrient transport and plant cooling.
- Water relations: Potassium provides much of the osmotic "pull" that draws water in to the plant roots from soil. K deficient plants are less able to with stand water stress, mostly because of their inability to fully utilize available water.
- Quality parameter: It increases the quality characters in plants due to its involvement in synthesis and transport of photosynthates (translocation of assimilates) to plant reproductive and storage organs (grain, fruit, tuber *etc.*)
- Enzyme activation: More than 60 enzymes have been identified that require potassium for their activation.
- It imparts winter hardiness.

- It reduces the lodging of crops.
- It counteracts harmful effects to excess nitrogen in plants.
- It imparts diseases resistance in plants.

Deficiency **Symptoms of potassium:** Deficiency symptoms of potassium are as follows:

- Deficiency symptoms of potassium develop first on older leaves of plant due to its high mobility in the plants.
- Chlorosis and necrosis of the leaf edges are observed.

    Chlorosis: *It is the loss of chlorophyll leading to yellowing in leaves. It is caused by the deficiency of elements like K, Mg, N, S, Fe, Mn, Zn and Mo.*

    Necrosis: *Necrosis means the death of tissues; particularly leaf tissue is caused by deficiency of K, Ca, and Mg.*
- Slow and stunted growth of plants occurs
- Another deficiency symptom is weakening of straw. Therefore, lodging occurs.
- Crop becomes susceptible to diseases attacks.
- Potassium deficient seeds and fruits are shriveled.

CALCIUM: Calcium (Ca) is a secondary nutrient element absorbed by the plant roots as calcium ions (Ca^{++}). Calcium in Ca - sufficient plants ranges from 0.2 to 1.0 per cent. Among all the nutrients, Ca is the most abundant in plant available forms in the soil. However, its deficiencies are more common in acid soils of the tropics (highly leached). Mass flow and root interception are the primarily mechanisms of Ca transport to the root surface.

Functions of calcium in plant:
- It is major constitute of calcium pectate of cell wall thus, it is essential to cell wall membrane structure and permeability.

    (Low Ca weakens cell membrane (i.e. plasma lemma), resulting in increased permeability, loss of cell contents, and failure of the nutrient uptake mechanisms.)
- Calcium and other cations neutralize organic acids formed during normal cell metabolism.
- It protects root cells against ion imbalance, low pH and toxic ions like Al.
- It is important to nitrogen metabolism and protein formation by enhancing NO_3^- uptake.
- Calcium is essential for cell elongation and cell division.
- It improves the uptake of other nutrients and encourages seed production.

Deficiency symptoms of calcium:

- Calcium is immobile in the plant and so it cannot move freely from older to younger parts of plants and that is why calcium deficiency symptoms are manifested at the tips of the shoots.
- Calcium deficiency causes deformed tissues and death of growing points including buds, blossoms and root tips.
- Leaf tips and margins are chlorotic and/or necrotic, a condition commonly referred to as die back or tip burn.
- Ca-deficient leaves become cup-shaped and crinkled.
- Root systems are stunted and root tips develop a dark colour and ultimately die.
- Blossom end rot in peppers tomatoes, deformed water melons, internal brown spot in tomatoes, discoloured and softer fruits and bitter pit in apples are the Ca deficiency symptoms.

MAGNESIUM: Like Ca, Mg occurs predominantly as exchangeable and solution Mg^{++}. Magnesium is absorbed by the plants as magnesium ions (Mg^{++}) ions. It is supplied to roots predominantly by mass flow. Its concentration in Mg - sufficient plants varies from 0.1 % to 0.4 %.

Functions magnesium in plants:

- Magnesium is a constituent of chlorophyll. Chlorophyll usually accounts for 15 to 20 per cent of total magnesium content of plant.
- It imparts dark green colour in leaves.
- Magnesium also serves as a structural component in ribosome.
- It is an activator of many enzyme systems involved in carbohydrate metabolism and synthesis of nucleic acids.
- Magnesium brings about significant increase in the oil content of several crops.

Deficiency Symptoms of magnesium:

- Because of mobility of Mg^{++} in plant and its ready translocation from older to younger leaves, thus, its deficiency symptoms often appear first on lower leaves.
- In many plants, Mg^{++} deficiency causes interveinal chlorosis in leaves, where only leaf veins remain green.
- Under severe Mg^{++} deficiency, leaf tissues become uniformly chlorotic or necrotic.

- Interveinal chlorosis with tints of red, orange and purple colours is observed in some of the vegetable crops.
- "Grass Tetany" (Hypomagnesaemia) is a nutritional disorder common in animals (cattle) grazing on Mg–deficient pastures well fertilized with K- fertilizer.

SULPHUR: Sulphur (S) is absorbed by the plants as the sulphate ions (SO_4^{-2}). Small quantity of SO_2 can be absorbed through plant leaves and utilized within plants, but high concentrations are toxic to plants. Like P and Mg, the concentration of S in healthy plants ranges from 0.1 % to 0.4 %. In a view of large field scale occurrence of deficiency in India, sulphur has been described as the fourth major nutrient after N, P, and K. Like N, the major faction of S in soils exists as organic S. Therefore, microbial transformation regulates the S availability in the soil.

Functions of sulphur in plants:
- Synthesis of some amino acids: Sulphur is required for the synthesis of the sulphur containing amino acids viz. Cystine, Cysteine and Methionine.
- Synthesis of chlorophyll: Although not a constituent, S is required for the synthesis of chlorophyll.
- Improve food quality: S-deficient plants accumulate non protein N as NH_2 and NO_3^-, in leaves which reduce food quality.
- Increases the oil quality and yield: Sulphur plays a major role in increasing the oil quality in oil seed crops. For examples: Oil palm, olive plant, ground nut, mustard *etc*.
- Increase the biological N fixation: Sulphur is a vital part of ferredoxins, which takes part in biological nitrogen fixation.
- It involved in the metabolic activities of vitamin, biotin, thiamine and coenzyme A.
- It increases root growth and stimulate the seed formation in plant

Deficiency symptoms of sulphur:
- Sulphur is less mobile in the plant, therefore their deficiency symptoms first observed on younger leaves.
- S deficiency has pronounced retarding effect on plant growth and is characterized by uniformly chlorotic (yellowing) leaves.
- In many plants, Sulphur deficiency symptoms resemble N deficiency and have undoubtedly led to many incorrect diagnoses. However, S is not easily translocated from older to younger plant parts, therefore, deficiency symptoms occur first in younger leaves.

FUNCTIONS, DEFICIENCY AND TOXICITY SYMPTOMS OF PLANT NUTRIENTS

IRON (Fe): Iron is taken up as ferrous ions (Fe^{+2}) by plants. Its concentration in the range of 100-500 mg/kg in mature leaf tissues is regarded sufficient for optimum crop production. Iron is a transition metal, exhibits two oxidation states - Fe (II) and Fe (III) - in plants and forms complexes with organic ligands. Variable valence of iron assigns it a role in biological redox systems.

Functions of iron in plants:
- Iron helps in the formation of chlorophyll
- Iron is a constituent of two groups of proteins: (i) Heme proteins and (ii) Fe-S proteins.
- It activates a number of enzymes
- It plays an essential role in the nucleic acid metabolism.
- It is necessary for synthesis and maintenance of chlorophyll in plants.

Deficiency Symptoms of iron:
- Deficiency of Fe results in interveinal chlorosis appearing first on the younger leaves with leaf margins and veins remaining green. Moreover, increased deficiency caused papery white colour of younger leaves.

    Interveinal chlorosis: Yellowing of leaves but, veins of leaves remain green coloured while, only areas between the veins of leaves become yellow in color. It is mainly caused due to deficiency of Fe, Mn and Zn.

- Under conditions of severe deficiency, growth cessation occurs with the whole plant turning necrotic.

MANGANESE: Manganese (Mn) is absorbed by the plants as manganous ions (Mn^{+2}). Healthy Mn-sufficient mature plants contains 20 to 300 ppm of Mn. Manganese a transition metal, is present in plants in Mn (II) form but, is easily oxidizable to Mn (III) and Mn (IV) forms.

Functions of manganese in plants
- The role of manganese is regarded as being closely associated with that of iron.
- Manganese helps in chlorophyll formation.
- Manganese is an integral component of the water splitting enzymes.
- Manganese also takes part in oxidation-reduction processes.
- It takes part in electron transport in photosynthesis II.

Deficiency Symptoms of manganese:
- Because of its essential role in photosynthesis, root and shoot growth rates are substantially reduced in Mn-deficient plants. Deficiency symptoms are more severe on middle leaves than on the younger leaves.

44 SOIL FERTILITY AND NUTRIENT MANAGEMENT

- Interveinal chlorosis on middle leaves. The interveinal chlorosis is characterized by the appearance of chlorotic and necrotic spots in the interveinal areas.
- In some plants (monocotyledonous), Mn deficiency symptoms appear as greenish grey spots, flecks, and stripes on the more basal leaves *(Grey speck)*.
- Chlorotic leaf areas soon become necrotic and turn red, reddish-brown or brown.

Toxicity symptoms of manganese
- Manganese toxicity occurs in Mn-sensitive crops grown in acid soils.
- Crinkle leaf in cotton is commonly observed due to Mn deficiency.

ZINC (Zn): Plants absorb Zn as zinc (Zn^{++}) ion. Zinc sufficient plant contains 27 to 150 ppm Zn in mature tissues. Since it does not have variable valence, it has no role in influencing redox processes directly.

Functions of zinc in plants
- Zinc is a constituent of enzymes *viz.* carbonic anhydrase, alcoholic dehydrogenase and superoxide dismutase.
- Zn is important in the synthesis of tryptophane (Tryptophane is component of some proteins and a compound needed for the production of growth hormones.
- Zinc influences the formation of some growth hormones such as indole acetic acid, metabolism of gibberellic acid and synthesis of RNA.
- Zn influences transport and translocation of P in plants.
- Zinc is helpful in reproduction of some plants.
- It is associated with the water uptake and water relations in the plant.
- Zn also involved in chlorophyll synthesis.

Deficiency symptoms of zinc
Among the micronutrient, the Zn deficiency is more common thought the world.
- Common deficiency of Zn is interveinal chlorosis, first appearing on the middle to young leaves.
- Light green, yellow, or white areas between leaf veins are observed.
- Zn deficiency reduced growth hormone production in plants which causes shortening of internodes and smaller leaves than normal one.
- Shortening of stem or stalk internodes, resulting in bushy, rosette leaves.
- Reduction in the size of young leaves, which are often clustered or borne very closely.

FUNCTIONS, DEFICIENCY AND TOXICITY SYMPTOMS OF PLANT NUTRIENTS

- In dicotyledonous plants, symptoms include short internodes (Rosetting) and decrease in leaf expansion (little leaf).
- *A Khaira disease of rice is* caused due to Zn deficiency. (*Khaira disease* inflicted rice fields gives a coppery brownish appearance and are characterized by missing hills).
- Rosetting commonly occurs in fruit and citrus trees.

Toxicity symptoms: Zn toxicity reduces or ceases root growth, resulting in yellowing of leaves and eventual plant death.

COPPER (Cu): Like other micronutrient cations, copper is absorbed by plant roots as cupric (Cu^{++}) ions. The concentration of copper in Cu-sufficient plants varies from 5 to 30 ppm. Copper is a transition element existing in the plant as a component of large number of proteins and enzymes.

Functions of Copper in plants

- Copper act as an electron carrier by changing its valence.
- Due to transition element, it involved in redox reactions in plant.
- Copper helps in utilization of iron during chlorophyll synthesis.
- Copper is constituent of number of enzymes.
- It is important in imparting disease resistance to the plants.
- It enhances the fertility of male flowers.

Deficiency Symptoms Copper

- Chlorosis in young leaves is common due to Cu deficiency.
- Male flowers sterility, delay flowering and senescence are the most important effects of copper deficiency.
- Chlorosis of younger shoot tissues, white tips, necrosis, leaf distortion and die back are the characteristic Cu-deficiency symptoms.

NICKEL (Ni): Nickel is the latest nutrient to be established as an essential nutrient to higher plants. It is taken up by the plant as nickel ion (Ni^{++}). Its concentration in Ni-sufficient plants varies from 0.1 to 1.0 ppm.

Functions of Nickel in plants

- Nickel is associated with nitrogen metabolism by way of influencing urease activity.

    In the systems were urea is used as the sole N fertilizer for foliar spray and Ni supply is poor, lower urease activity causes urea toxicity to the foliage and leads to severe necrosis of the root tips.

- It facilitates transport of nutrients to the seeds or grains.
- In free living *Rhizobia*, adequate Ni supply ensures optimum hydrogenase activity.

Deficiency Symptoms of Nickel

- Nickel (Ni) deficient plants accumulate toxic levels of urea in leaf tips because of reduced urease enzyme activity.
- Ni deficient plant may develop chlorosis in the youngest leaves that progresses to necrosis of the meristems.

BORON: Boron is absorbed by the plant mainly as boric acid (H_3BO_3). However, it can be absorbed in some of its ionic forms *viz.* dihydrogen borate ($H_2BO_3^-$), monohydrogen borate (HBO_3^{-2}) and borate (BO_3^{-3}) under high pH conditions. Normal boron sufficient plants have B-contents ranges from 10 to 200 ppm. Boron is neither a constituent of enzymes nor it activates any of the enzymes.

Functions of Boron in plants: Followings are the major functions of Boron (B) in plants:

- Boron is responsible for cell wall formation and stabilization, lignification and xylem differentiation.
- It imparts the drought tolerance to the crops.
- Boron plays role in pollen germination and pollen tube growth.
- Boron increases the solubility of calcium as well as mobility of calcium in the plant.
- It acts as K/Ca ratio in the plant.
- Boron is required for the synthesis of amino acids and proteins.

Deficiency symptoms

- Boron deficiency symptoms become conspicuous on the terminal buds or youngest leaves, which become discoloured and may die under acute conditions of B deficiency.

(As both, B and Ca are immobile elements in the plant; they are not easily translocated from older leaves to growing tips of plant as well in reproductive organs of plant. Therefore, their deficiency symptoms are first observed at terminal buds (growing tips) of plant. Fruit cracking is also associated with Ca and B deficiency in plants due to above reasons.)

FUNCTIONS, DEFICIENCY AND TOXICITY SYMPTOMS OF PLANT NUTRIENTS

- B-Deficiency often appears in the form of thickened, wilted or curled leaves and cracking or rotting of fruits, tubers or roots.
- Internodes become shorter and give bushy or rosette appearance.
- Increased diameter of stem and the petioles give rise to the typical cracked stem of celery.
- Typical names given to B-deficiency in different crops are:
  - Heart rot of sugar beet and man gold
  - Browning or hollow stem of cauliflower
  - Top sickness of tobacco
  - Internal cork of apple

MOLYBDENUM (Mo): Molybdenum is the only heavy transition metal taken up by the plants as molybdate ion (MoO_4^{-2}). A healthy Mo-sufficient plant contains 0.1 to 2 ppm of Mo. Ability of Mo to exist in variable valence states *imparts it a biochemical role.*

Functions of Molybdenum in plants:

- Molybdenum is an essential component of NO_3 reductase enzyme.
- Molybdenum (Mo) is essential for biological N fixation: It is a structural component of nitrogenase enzyme which actively involved in nitrogen fixation by root nodule bacteria of leguminous crops, by some algae and actinomycetes, and by free living nitrogen fixing organisms such as *Azotobacter.*
- Molybdenum affects the formation and viability of pollens and development of anthers.

Deficiency symptoms of Molybdenum:

- The critical concentration of molybdenum deficiency in plants is usually less than 0.1 ppm.
- Molybdenum deficiencies resemble the N and Fe deficiencies.

CHLORIDE (Cl): Nearly all chloride in soils exists in soil solution as Cl^- ion, and it is also taken up by the plants in Cl^- form by plant roots. The concentration of Cl in plants can be similar to S but, being a micronutrient the optimum requirement is less *(Av.* 100 ppm) and high levels often toxic to plants.

Functions of Chloride in plants

- Chloride principally is involved in osmotic and ion charge balance, which are important to many biochemical processes in plants.

- Over 100 chloride containing organic compounds are known in plants; however their functions are not well understood.
- It plays an important role in photosynthesis process by exchanging the gases (CO_2 and O_2).
- It is also essential for maintaining electrical balance in tonoplasts.

Deficiency symptoms of Chloride
- Chlorosis in younger leaves and overall wilting of plants are the two most common deficiency symptoms of chloride.
- Necrosis in some plant parts, leaf bronzing, and reduction in root and leaf growth may also be observed.

Toxicity symptoms
- Due to excess chloride concentration in plant, leaves become thickened and tend to roll.
- Storage quality of tuber crops is reduced.
- Excess concentration in soil increased the osmotic pressure and thus reduces the water and nutrient uptake by plant.

3.2. BENEFICIAL ELEMENTS

Beneficial elements are the mineral elements which are stimulate plant growth, but are not essential or which are essential only for certain plant species, or under specific conditions. Sodium, Silicon, cobalt, selenium and vanadium are the beneficial elements.

SN	Element	Important functions
1.	Sodium (Na)	Sodium can replace some of the essential functions of potassium in plants like sugar beet, turnip, table beet etc. It improves water balance of plants under limited water supply or under arid climate.
2.	Silicon (Si)	Silicon has a beneficial role in rice and sugarcane crops.
3.	Cobalt (Co)	Enhances the symbiotic nitrogen (N) fixation in legumes and improve the nutritional quality of forage crops for ruminants.
4.	Vanadium (V)	It is an essential element for some microorganisms. It favours nitrogen fixation. The chemical behaviour of V and Mo is similar.
5.	Selenium (Se)	Essential for animals

FUNCTIONS, DEFICIENCY AND TOXICITY SYMPTOMS OF PLANT NUTRIENTS

3.3. PLANT LOCATION SPECIFIC DIAGNOSTIC KEYS

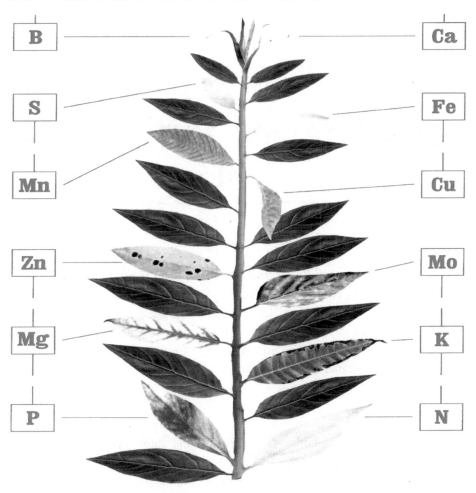

VISUAL DIAGNOSTIC KEYS FOR INDENTIFICATION

Mobile Elements (symptoms in older leaves)	Immobile Elements (symptoms in younger leaves)
N : Total chlorosis of leaves	Ca : Crinkling and drying of terminal leaves
P : Purple coloration of leaves	S : Total chlorosis in terminal leaves
K : Marginal drying of leaves	Fe : Interveinal chlorosis of young leaves
Mg : Reddening of leaves	Mn : Interveinal chlorosis with necrosis
Zn : Bronzing of older leaves	Cu : Dark green leaves with white blotches
Mo : White streaks and lean leaves	B : Fruit cracking, death of apical shoots

CHAPTER 4

CHEMISTRY OF NUTRIENT IN SOILS

The changes undergone by common fertilizers after these are taken out of the bag and added to soils are discussed. By understanding the fate of fertilizers, measures for increasing their efficiency can be suggested and adopted. When fertilizers react with soils, the compounds produced are by and large similar to the ones which are present in soils and which are produced by the breakdown of minerals and organic matter. That is why soils accept fertilizers without any fuss.

4.1. NITROGEN

Nitrogen occurs in soil in both cationic (NH_4^+) and anionic (NO_3^-, NO_2^-) forms, the greater parts occurs in organic forms. NH_4^+ fixed on the cation exchange sites, is tightly bound by clay and is slowly available to plants. The available nitrates and ammonium form is only 1-2% of the total soil nitrogen. Nitrate is highly mobile. Nitrogen availability depends upon the rate at which organic nitrogen is converted to inorganic nitrogen (mineralization). Most soil nitrogen is unavailable to plants. The amount in available forms is small and crops withdraw a large amount of nitrogen. Two forms of nitrogen available to plants are nitrate (NO_3^-) and ammonium (NH_4^+). Roots can absorb both of these forms, although many species preferentially absorb nitrate-nitrogen over ammonium-nitrogen.

Nitrogen transformation in Soils

The cycling of N in the soil-plant-atmosphere system involves many transformations of N between inorganic and organic forms. Nitrogen is subjected to amino compounds (R-NH$_2$, R represents the part of the organic molecules with which amino group (NH$_2$) is associated), then to ammonium (NH$_4^+$) ion and nitrate (NO$_3^-$). Ammonium nitrogen is often converted to nitrate-nitrogen by micro-organisms before absorption through a process called nitrification.

Nitrogen Mineralization

The conversion of organic N to NH$_4^+$ and NO$_3^-$ is known as nitrogen mineralization. Mineralization of organic N involves two reactions, aminisation and ammonification, which occur through the activity of heterotrophic micro-organisms. The enzymatic process may be indicated as follows:

$$RNH_2 \underset{-H_2O}{\overset{+H_2O \quad Mineralization}{\rightleftharpoons}} ROH + NH_4 \underset{-O_2}{\overset{+O_2}{\rightleftharpoons}} NO_2^- + 4H \underset{-O}{\overset{+O}{\rightleftharpoons}}$$

Immobilization

Aminisation

The decomposition of protein into amines, amino acids and urea is known as aminisation.

$$\text{Proteins} \xrightarrow[\text{Bacteria, H}]{H_2O} \underset{\text{Amino acids}}{R - \underset{|}{\overset{NH_2O}{C}} - COOH} + \underset{\text{Amines}}{R-NH_2} + \underset{\text{Urea}}{H_2N - \overset{O}{\overset{\|}{C}} - NH_2} + CO_2 + \text{Energy}$$

Fungi

Ammonification

The step, in which, the amines and amino acids produced by aminisation of organic N are decomposed by other heterotrophs, with the release of NH$_4^+$, is termed as ammonification.

$$R-NH_2 + H_2O \rightarrow NH_3^+ \quad R-OH + \text{Energy}$$
$$\downarrow$$
$$H_2O$$
$$NH_4^+ + OH^-$$

Nitrogen immobilization

Immobilization is the process in which available forms of inorganic nitrogen (NO_3^- NH_4^+) are converted to unavailable organic nitrogen. Immobilization includes assimilation and protein production so those inorganic ions are made into building block of large organic molecules.

Nitrification

Nitrification is a process in which NH_4^+ released during mineralization of organic N is conveted to NO_3^-. it is a two step process in which NH_4^+ is converted first to NO_2^- and then to NO_3^-. Biological oxidation of NH_4^+ to NO_2^- is represented by:

$$2NH_4^+ + 3O_2 \xrightarrow{\text{Nitrosomonas}} 2NO_2^- + 2H_2O + 4H^+$$

NO_2^- is further oxidized to NO_3^- be bacteria

$$2NO_2^- + O_2 \xrightarrow{\text{Nitrobactor}} 2NO_3^-$$

4.2. PHOSPHORUS

Organic and inorganic forms of phosphorus occur in soils and both the forms are important to plants as source of phosphorus. The relative amounts of phosphorus in organic and inorganic forms vary greatly from soil to soil.

Organic phosphorus compounds

Organic phosphorus represents about 50% of the total P in soils (Varies between 15 and 80% in most soil. Most organic P compounds are esters of orthophosphoric acid and have been identifies primarily as (a) inositol phosphates, (b) phospholipids and (c) nucleic acids.

Inorganic phosphorus compounds

Most inorganic phosphorus compounds in soil fall into one of the two group: (a) those in which calcium is the dominant controlling cation (calcium phosphate) and (b) those in which iron and aluminum are the controlling cations (iron and aluminum phosphates).

Phosphate Retention and Fixation

Phosphate anions can be attracted to soil constituents with such a bond that they become insoluble and not easily available to plants. This process is called phosphate fixation or retention.

Phosphate retention

Acid soils usually contain significant amounts of soluble and exchangeable Al_3^+, Fe_3^+ and Mn_2^+ ions. Phosphate, when present, may be adsorbed to the colloid surface with these ions serving as a bridge. This phenomenon is called co-adsorption. The phosphate retained in this way is still available to plants. Such a reaction can also take place with Ca-saturated clays.

Ca clay adsorbs large amounts of phosphate. The Ca^{2+} ions forms the linkage between the clay and phosphate ions as: Clay-Ca-H_2PO_4.

The phosphate ions can also enter into a chemical reaction with the foregoing free metal ions as: $Al^{3+} + 3H_2PO_4^- \rightarrow Al(H_2PO_4)_3\downarrow$. The product formed is not soluble in water and precipitates from solution. With the passage of time the Al phosphate precipitates, become less soluble and less available to plants. The lower the soil pH, the greater the concentration of soluble Fe, Al, and Mn: consequently, larger the amount of phosphorus retention in this way.

Phosphate fixation in acidic soils

Many acidic soils contain high amounts of free Fe and Al and Fe and Al hydrous oxide clays. The free Fe, Al and the sesquioxide clays react rapidly with phosphate, forming a series of not easily soluble hydroxyl phosphates.

The amount of phosphate fixed by this reaction usually exceeds that fixed by phosphate retention. Generally, clays with low sesquioxide ratios (SiO_2/R_2O_3) have a higher P-fixing capacity.

Phosphate fixation in alkaline soils

Many alkaline soils contain high amounts of soluble and exchangeable Ca2+ and, frequently, CaCO3. Phosphate react with both the ionic and carbonate form of Ca.

$$3Ca^{2+} + 2PO_4^{3-} \longrightarrow Ca_3(PO_4)_2 \text{ (Insoluble)}$$

$$3CaCO_3 + 2PO_4^{3-} \longrightarrow Ca_3(PO_4)_2 + 3CO_2 \text{ (Insoluble)}$$

Phosphate fixation cannot be avoided entirely, but it may be reduced by addition of competing ions for fixing sites. Organic anions from stable manure and silicates are reported to be very useful in reducing P fixation.

4.3. POTASSIUM

Forms and availability of potassium in soils

Potassium in soil occurs in four phases namely soil solution phase, exchangeable phase, non-exchangeable phase and mineral phase. The different forms are in dynamic equilibrium with one another.

The forms of potassium in soils were positively and significantly correlated with K content in silt and clay.

Water soluble K

The water soluble K is the fraction of soil potassium that can be readily adsorbed by the growing plants. However this is a very small fraction of total K. The dilution of the soil increases the concentration of water-soluble K and drying decreases it further. It is about 1 to 10 mg kg^{-1} of total K.

Exchangeable K

Exchangeable K is held around negatively charged soil colloids by electrostatic attraction. Thus, exchangeable potassium represents that fraction of K, which is adsorbed on external and accessible internal surfaces. It is about 40 to 60 mg kg^{-1} of total K.

Non-exchangeable (fixed) K

Potassium held at inter lattice position is generally non-exchangeable. Non-exchangeable K is distinct from mineral K in that it is not bonded covalently

CHEMISTRY OF NUTRIENT IN SOILS

within the crystal structures of soils mineral particles. Instead, it is held between adjacent tetrahedral layers of dioctahedral and trioctahedral micas, vermiculites and intergraded clay minerals. It is about 50 to 750 mg kg-1 of total K.

Mineral (lattice) K

Lattice K is a part of the mineral structure and is available to the plants very slowly. (As compared to the non-exchangeable K). Both the rate and amount of lattice K released to plants depend on the quantity of clay, especially the smaller clay particles, and its mineralogy. It is about 5,000 to 25,000mg kg^{-1}.

For convenience, the various forms of potassium in soils can be classified on the basis of availability in three general groups: (a) unavailable (b) readily available and (c) slowly available.

A dynamic equilibrium of various forms of K in the soil may be shown as :

$$K(lattice) \leftrightarrow K(exchangeable) \leftrightarrow K (solution)$$

Relatively Unavailable Forms

The greatest part (90-98%) of all soil potassium in a mineral soil is in relatively unavailable forms. The compounds containing most of this form of potassium are the feldspars and micas. These minerals are quite resistant to weathering and probably supply relatively insignificant quantities of potassium during a given growing season.

Readily Available Forms

The readily available potassium constitutes only about 1-2% of the total amount of this element in an average mineral soil. It exists in soils in two forms; (i) potassium in soil solution and (ii) exchangeable potassium adsorbed on soil colloidal surfaces. Most of this available potassium is in the exchangeable form (approximately 90%). Soil solution potassium is most readily adsorbed by higher plant and is, of course, subject to considerable leaching loss.

Slowly Available Forms

In the presence of vermiculite, smectite, and other 2:1- type minerals the potassium of such fertilizers as muriate of potash not only becomes adsorbed but may become definitely 'fixed' by the soil colloids. The potassium as well as ammonium ions fit in between layers in the crystals of these normally expanding clays and become an integral part of the crystal. Potassium in this form cannot be replaced by ordinary exchange methods and consequently is referred to as non-exchangeable potassium. As such this element is not readily available to higher plants. This form is in equilibrium, however, with the available forms and consequently acts as an extremely important reservoir of slowly available potassium.

4.4. SULPHUR TRANSFORMATION IN SOIL

Sergei Nikolaievich Winogradsky (1856 – 1953) was microbiologist, ecologist and soil scientist who pioneer for his notable work on bacterial sulfate reduction. The transformation of sulphur is important indicators of its availability to plants. Availability of sulphur from organic sulphur reserves in soils depends on its mineralization through microbial activity.

Sulphur Oxidation

Sulphur oxidation occurring in soils is mostly biochemical in nature. Sulphur oxidation is accomplished by number of autotrophic bacteria including those of genus *Thiobacillus*, five species of which have been characterized:

(a) *Thiobacillus thioxidans* (b) *T. thiparus* (c) *T. nonellus* (d) *T. denitrificans* (e) *T. ferooxidans*

In soils, sulfides, elemental sulphur, thiosulphates and polythionates are oxidized.

Oxidation reactions

$$H_2S + 2O_2 \rightarrow H_2SO_4 \rightarrow 2H^+ + SO_4^{2-}$$
$$2S + 3O_2 + 2H_2O \rightarrow 2H_2SO_4 \rightarrow 4H^+ + 2SO_4^{2-}$$

Thus S-oxidation is an acidifying process.

Sulphur reduction

Sulphate tends to be unstable in anaerobic environments so they are reduced sulfides by a number of bacteria of two genera, *Desulfovibro* (five species) and *Desulfotomaculum* (three species).

The organisms use the combined oxygen in sulfate to oxidize organic materials.

Reduction reactions

$$2R\text{-}CH_2OH + SO_4^{2-} \longrightarrow 2R\text{-}COOH + 2H_2O + S^{2-}$$
(Organic alcohol) (Sulfate)   (Organic acid)   (Sulfide)

Also, sulfites (SO_3^{2-}), thiosulfates ($S_2O_3^{2-}$) and elemental sulphur (S) are rather easily reduced to the sulfides form by bacteria and other organisms.

The oxidation and reduction of inorganic sulphur compounds are of great importance to growing plants. These reactions determine the quantity of sulfate present in soils at any one time. Also, the state of sulphur oxidation determines to a marked degree the soil acidity as S-oxidation is an acidifying process.

4.5. CALCIUM AND MAGNESIUM TRANSFORMATIONS IN SOIL

Calcium is an important amendment element in saline and alkali soils. Calcium application helps in correcting the toxicity and deficiency of several other nutrients. The main transformations of Ca and Mg in soils are (i) solubilization and leaching and (ii) conversion into less soluble fractions by adsorption.

Solubilization and leaching of calcium and magnesium: It is affected by following:

Soil texture: Losses are more in light textured soils because of high permeability and percolation of rain and irrigation water.

Rainfall: As the rainfall increases the loss of Mg and Ca also increases.

Organic matter: Application of organic matter leads to net loss of Ca and Mg from the soil.

Ferrolysis: High amounts of bases such as Ca^{2+} and Mg^{2+} may be lost from the exchange complex and leached by high amounts of cations such as Fe^{2+} and Mn^{2+} which are released following reduction of soil. This is called **Ferrolysis**.

Conversion of calcium and magnesium into less soluble form by adsorption:

Calcium and magnesium in soil solution and in exchange complex are in a state of dynamic equilibrium. When their concentration in solution decreases, Ca and Mg coming from the exchange complex replenish this. On the other hand if their concentration in soil solution is high, there is tendency towards their being adsorbed on the exchange complex.

4.6. FE AND ZN TRANSFORMATIONS IN SOIL

A. IRON

The most important chemical change that takes place when a soil is submerged is the reduction of iron and the accompanying increase in its solubility. The intensity of reduction depends upon time of submergence, amount of organic matter, active iron, active manganese, nitrate etc.

Due to reduction of Fe^{3+} to Fe^{2+} on submergence, the colour of soil changes from brown to grey and large amounts of Fe^{2-} enter into the soil solution. It is evident that the concentration of ferrous iron (Fe^{2+}) increases initially to some peak value the thereafter decreases slowly with the period of soil submergence. Organic matter also enhances the rate of reduction of iron in submerged soils. The initial increase in the concentration of ferrous iron (Fe2+) on soil submergence is caused by the reduction that are shown below:

$$\underset{(Insoluble)}{Fe(OH)_3} + e^- \underset{}{\overset{Reduction}{\rightleftharpoons}} \underset{(Soluble)}{Fe^{2+}} + 3OH^-$$

The decrease in the concentration of Fe^{2+} following the peak rise is caused by the precipitation of Fe^{2+} as $FeCO_3$ in the early stages where high partial pressure of CO_2 prevails and *as* $Fe_3(OH)_8$ due to decrease in the partial pressure of $CO_2 (pCO_2)$

$$\underset{(Soluble)}{2Fe^{2+}} + 3CO_2 \longrightarrow \underset{(Insoluble)}{2FeCO_3}$$

Rice benefits from the increase in availability of iron but may suffer in acid soils, from an excess. The reduction of iron has some important consequences: (i) the concentration of water soluble iron increases, (ii) pH increases, (iii) cations are displaced from exchange sites, *(iv)* the solubility of P and Si increases and *(v)* new minerals are formed.

A schematic representation for the transformation of iron in submerged soils is shown below:

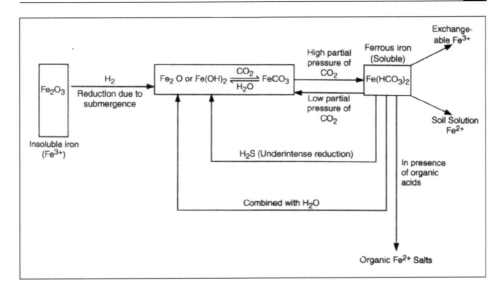

B. ZINC

The transformation of zinc in submerged soils is not involved in the oxidation-reduction process like that of iron and manganese. However, the reduction of hydrous oxides of iron and manganese, changes in soil pH, partial pressure of CO_2, formation insoluble sulphide compound etc. In soil on submergence is likely to influence the solubility of Zn in soil either favorably or adversely and consequently the Zn nutrition of low and rice. The reduction of hydrous oxides of iron and manganese, formation of organic complexing agents, and the decrease in pH of alkaline and calcareous soils on submergence are found to favor the solubility of Zn, whereas the formation of hydroxides, carbonates, sulphides may lower the solubility of Zn in submerged soils. Zinc deficiency in submerged rice soils is very common owing to the combined effect of increased pH, HCO_3^- and S^{2-} formation.

The solubility of native forms of Zn in soils is highly pH dependent and decreases by a factor of 10^2 for each unit increase in soil pH. The activity of Zn-pH relationship has been defined as follow:

$$\text{Soil} + Zn^{2+} \rightleftharpoons \text{Soil} - Zn + 2H^+$$

The pK value for the above reaction with the solid phase of soils is 6.0. This equation holds good for submerged soils. Some equations relating to solubility of Zn in submerged soils governed by various metastable compounds are given below:

SOIL FERTILITY AND NUTRIENT MANAGEMENT

$3Zn^{2+} + 2PO_4^{3-}$	$\rightleftharpoons$	$Zn_3(PO_4)_2$	$pK_2 = 32.0$
$Zn^{2+} + NH_4^+ + PO_4^{3-}$	$\rightleftharpoons$	$ZnNH_4PO_4$	$pK_s = 16.0$
$Si(OH)_4 + Zn^{2+}$	$\rightleftharpoons$	$Zn(OH)_2 + SiO_2 + 2H^+$	$pK_2 \geq 17.5$
$Zn^{2+} + S^{2-}$	$\rightleftharpoons$	ZnS	$pK_2 = 22.8$
$Zn^{2+} + 2OH^-$	$\rightleftharpoons$	$Zn(OH)_2$	$pK_s = 17.5$
$Zn^{2+} + CO_3^{2-}$	$\rightleftharpoons$	$ZnCO_3$	$pK_s = 16.5$

Many of these compounds are metastable intermediate reaction products and varying mean residence time in submerged soils. Applied Zn tends to approach the solubility of the native forms instead of having residual effect in the former Zn forms.

When an aerobic soil is submerged, the availability of native as well as applied Zn decreases and the magnitude of such decrease vary with the soil properties. The transformation of Zn in soils was found to be greatly influenced by the depth of submerged and application of organic matter. If an acid soil is submerged, the pH of the soil will increase and thereby the availability of Zn will decrease. On the other hand, if an alkali soil is submerged, the pH of the soil will decrease and as a result the solubility of Zn will generally increase.

The availability of Zn decreases due to submergence may be attributed to the following reasons:

(i) Formation of insoluble franklinite ($ZnFe_2O_4$) compound in submerged soils.

$$Zn^{2+} + 2Fe^{2+} + 4H_2O \underset{}{\overset{reduction}{\rightleftharpoons}} ZnFe_2O_4 \, (Franklinite)$$

(ii) Formation of very insoluble compounds of Zn as ZnS under intense reducing conditions.

$$Zn^{2+} + S^{2-} \rightleftharpoons ZnS \, (Sphalerite)$$
(Very insoluble in water)

(iii) Formation of insoluble compounds of Zn as $ZnCO_3$ at the later period of soil submergence owing to high partial pressure of $CO_2 (PCO_2)$ arising from the decomposition of organic matter.

$$Zn^{2+} + CO_2 \rightleftharpoons ZnCO_3 \, (Smithsonite)$$
(Insoluble)

CHEMISTRY OF NUTRIENT IN SOILS

(iv) Formation of Zn(OH)2 at a relatively higher pH which decreases the availability of Zn.

$$Zn^{2+} + 2HO^- \rightleftharpoons Zn(OH)_2 \text{ (Insoluble)}$$

(v) Adsorption of soluble Zn^{2+} by oxide minerals *e.g.* sesquioxides, carbonates, soil organic matter and clay minerals etc. decreases the availability of Zn, the possible mechanism of Zn adsorption by oxide minerals is shown below :

Mechanism I

In mechanism I, Zn^{2+} adsorption occurs as bridging between two neutral sites, but in addition to this mechanism, Zn^{2+} could also be adsorbed to two positive sites or to a positive and neutral site.

Mechanism II

This mechanism occurs at low pH and results non-specific adsorption of Zn^{2+}. In this way Zn^{2+} is retained and rendered unavailable to plants.

$$Zn^{2+} + PO_4^{3-} \rightleftharpoons Zn_3(PO_4)_2 \text{ (Insoluble)}$$

(vi) Formation of various other insoluble zinc compounds which decreases the availability of Zn in submerged soils, *e.g.* high Phosphatic fertilizer induces the decreased availability of Zn^{2+}.

SOIL FERTILITY AND NUTRIENT MANAGEMENT

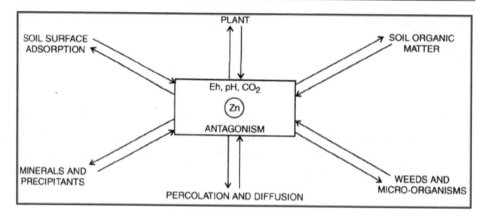

A simplified diagram illustrating dynamic equilibria of Zn in submerged soils is shown in figure.

It shows that rice receives Zn from the soil solution and the exchangeable and adsorbed solid phase including the soil organic fractions.

Zinc sulphide (ZnS, Sphalerite) in the presence of traces of hydrogen sulphide (H_2S) in submerged soils may control the solubility of Zn. Zinc is stable in submerged soils. So it can be concluded that higher the pH and poorer the aeration, the greater is the likelihood of Zn deficiency if the soil solution Zn activity is controlled by sphalerite (ZnS).

Therefore, a variety of chemical reactions in soils influence the availability of Zn to rice. For example, high manganese concentration antagonizes Zn absorption and translocation.

4.7. Q/I RELATIONSHIP

In addition to these, the availability of Zn in submerged soils is governed by the mutual interaction of quantity *(q)* intensity *(c)*, and kinetic parameters as regulated by the adsorption, desorption, chelation and diffusion of Zn from soils to the plant roots. The quantity-intensity relationship of Zn in submerged soils may be described by the linear form of the Langmuir type equation. The supply parameter assumes the form,

$$\text{Supply parameter} = qc^{1/2} \cdot K_1 K_2^{-1/4} \text{ or } \sqrt{cq/\sqrt{K_1 K_2}}$$

where q is the quantity c is the intensity, K_1 and K_2 are constants.

The optimum Zn supply to rice is ensured when the value of the supply parameter is unity (1.0).

Different crop management factors combined influence the availability of Zn to rice like, native Zn content of the soil, soil pH, organic matter, submergence, partial pressure of CO_2, HCO_3, organic acids, various natural interactions, environmental effects and water quality etc.

CHAPTER 5

SOIL FERTILITY EVALUATION

Optimum plant health and productivity of crops depends on adequate supply of nutrients. The quantity of nutrients required by crops depending on many interactive factors such as: (i.) plant species and variety, (ii.) yield potential of grown crop and variety, (iii.) soil properties (iv.) environmental factors and (v) management practices following by producer.

When soil does not supply sufficient nutrients in balanced amount for optimum growth and production, nutrients must be applied from outer sources in the forms of manures and fertilizers. The proper amount of nutrients requirement to a crop knows by (i) quantity of nutrients required by crop for potential yield and (ii) the nutrient supplying power of soils. Hence, the evaluation of soil fertility becomes very important for nutrient management. Soil fertility evaluation means to assess the nutrients supplying capacity of soil which is essential for balanced nutrition of the crops. Balance nutrients use refers to the application of plant nutrients in right amount and proportion using correct methods of fertilizers application and time of application. It helps in maintenance and improving soil productivity. Several techniques are commonly employed to assess the fertility status of the soils. A proper evaluation of the fertility of a soil before planting of a crop helps in adopting appropriate measures to make up for the shortcoming and ensuring a good crop production.

5.1. METHODS OF SOIL FERTILITY EVALUATION

Following methods are widely used to assess the soil fertility:

1. **Nutrient deficiency symptoms of plant:** Identifying the typical nutrient deficiency symptoms on plant under the field.

2. **Plant tissue analysis:** Analyzing the tissues of growing plants in the laboratory.
3. **Sensor based plant tissue Analysis:** Measuring the absorbance or reflectance of light under field condition It is an indirect estimate of nutrient concentration in plant.
4. **Biological tests:** The growth of higher plants or certain micro-organisms is used as a measure of soil fertility.
5. **Soil testing:** Representative soil samples are analyzed for evaluating soil fertility in the laboratory

(1) NUTRIENT DEFICIENCY SYMPTOMS OF PLANT: As we know that the plant requires seventeen essential nutrients for their optimum growth and development. When the availability/uptake of nutrient is very less than the plant requirement for optimum growth, it shows deficiency symptoms. These symptoms are nutrient specific and show different patterns in crops for different essential nutrients (as discussed in nutrient deficiency symptoms caption). It is good tool to detect deficiencies of nutrient in the field but, these techniques have several limitations which are as follows:

- Similar visual symptoms may be caused by more than one nutrient. For example, chlorosis is caused due to deficiency of N, S, Mg, Fe and Zn.
- Deficiency of one nutrient may be related to an excess quantity of another.
- It is difficult to distinguish among the deficiency symptoms in the field, as disease or insect damage can resemble certain micronutrient deficiencies.
- Nutrient deficiency symptoms are observed only after the crop has already suffered an irreversible loss.

There are following indicator plants which show the nutrient deficiencies or excesses.

Plant	:	Nutrient deficiency/toxicity
Oat	:	Mg, Mn and Cu deficiencies
Wheat and barley	:	Mg, Cu and sometimes Mn deficiencies
Sugar beets	:	B and Mn deficiencies
Maize	:	N, P, K, Mg, Fe, Mn and Zn deficiencies
Potatoes	:	K, Mg and Mn deficiencies
Rape seed	:	N, P and Mg deficiencies
Brassica species	:	K and Mg deficiencies
Cauliflower	:	B and Mo deficiencies
Barley	:	B, Mn and Al toxicities
Cucumber	:	N and P excess
Sugar beets	:	Cu excess

(2) PLANT TISSUE ANALYSIS: Plant analysis in a narrow sense is the determination of the concentration of nutrient or extractable fraction of nutrient from a plant sample taken from a particular part or portion of a crop at a certain time/stage of morphological development.

Nutrient content: Nutrient content is defined as the concentration of nutrient per unit dry biomass of plant. It is expressed in terms of percentage or ppm.

Plant analysis is complementary to soil testing. As, in many situations, available nutrient content of soil fails to correlate with nutrient content in plant or the growth and yield of crop. This can be attributed to many reasons such as (i) poor physical condition of soil, (ii) some problems associated with soil *viz.* soil acidity, salinity *etc.*, (iii) root growth pattern and (iv.) environmental conditions. On the other hand, the nutrient content in the plant tissue is, generally, positively correlated with plant growth. Therefore, plant analysis has been used as a diagnostic tool to determine the nutritional cause of plant disorders/diseases.

Types of plant analysis: Two types of plant analysis are generally followed to estimate nutrient content in plant:

(a) *Plant tissue analysis:* In this test, fresh plant tissues are analyzed in the laboratory by using standard methods. Plant tissue analysis includes following steps:

1. Collection of the representative plant parts from the field at the specific growth stage.
2. Washing, drying and grinding of plant tissue in laboratory.
3. Oxidation of the powdered plant samples.
4. Estimation of different elements by using standard methods and
5. Interpretation of the status of nutrients with respect to deficiency/sufficiency/toxicity on the basis of known critical concentrations.

Plant Testing:

1. Analysis of tissues from plant growing on the soil: Plant analysis in a narrow sense is the determination of the concentration of an element or extractable fraction of an element in a sample taken from a particular part or portion of a crop at a certain time or stage of morphological development

Plant analysis is complementary to soil testing. In many situations, the total or even the available content of an element in soil fails to correlate with the plant tissue concentration or the growth and yield of crop. This can be ascribed to many reasons including the physico chemical properties of the soils and the root growth patterns. On the other hand, the concentration of an element in the plant tissue is, generally, positively correlated with the plant health. Therefore, the plant analysis has been used as a diagnostic tool to determine the nutritional cause of plant disorders/diseases. The plant analysis constitutes (1) the collection of the representative plant parts at the specific growth stage, (2) washing, drying and grinding of plant tissue, (3) oxidation of the powdered plant samples to solubilize the elements, (4) estimation of different elements, and (5) interpretation of the status of nutrients with respect to deficiency / sufficiency /toxicity on the basis of known critical concentrations.

2. **Collection and Preparation of plant samples:** Plant scientists have been able to standardize the procedures for collection of samples of plant tissue with respect to the plant part and growth stage, which reflect the nutrient concentrations corresponding to the health of the growth because the concentrations of different nutrients vary significantly over the life cycle of a plant. Generally, the recently matured fully expanded leaves just before the onset of the reproductive stage are collected and put in perforated paper bags. The plant samples are often contaminated with dust, dirt and residues of the sprays, *etc.* and need to be washed first under a running tap water followed by rinsing with dilute HCl (0.001N), distilled water and finally in deionized water. The washed samples are dried in a hot air oven at 60±5°C for a period of 48 hours and ground in a stainless steel mill to pass through a sieve of 40/60 mesh.

3. **Oxidation of plant material:** The main objective of oxidation is to destroy the organic components in the plant material to release the elements from their combinations. The plant materials can be oxidized by either dry ashing at a controlled high temperature in a muffle furnace or wet digestion in an acid or a mixture of two or more acids.

    (i) ***Dry-ashing*:** The powdered plant materials in tall form silica crucibles are ashed at 500°C in a muffles furnace for 3-4 hours. High temperatures are likely to result in the loss of some volatile elements but with adjusting the time of muffling between 2-72 hours, any significant effect on the analytical results can be avoided. Nitrogen and sulphur, being highly volatile, are lost more or less completely during dry ashing even at 500°C but at higher temperatures, elements like K are also reported to be lost. Thus, temperature is an important consideration in dry ashing. The ash is dissolved in 2ml of 6N HCl, heated on a hot plate to near dryness and taken in 10 ml dilute HCl (0.01N) or 20% aqua regia before making up the final volume with distilled water. These extracts contain different amounts of insoluble materials, mainly silica, depending upon the plant species. These insoluble materials settle down on keeping for sometime or can be separated by filtration before estimation of different elements. All elements, except N and S, can be estimated in these extracts by any technique. In general, the results obtained by this method, are quite satisfactory and comparable to those obtained by this method, are quite satisfactory and comparable to those obtained by wet digestion procedures. Moreover, B can only be determined by dry ashing since it is volatilized during wet digestion with di-or triacid mixtures.

(ii) Wet Digestion: Wet oxidation digestion reagents and their applicability

Sr. No.	Reagents	Applicability to organic manure	Remarks
1	H_2SO_4/HNO_3	Vegetable origin	Most commonly used
2	H_2SO_4/H_2O_2	Vegetable origin	Not very common
3	HNO_3	Biological origin	Easily purified reagent, short digestion time, temperature 350°C
4	$H_2SO_4/HClO_4$	Biological origin	Suitable only for small samples, danger of explosion
5	$HNO_3/HClO_4$	Protein, carbohydrate (no fat)	Less explosive
6	$HNO_3/HClO_4/H_2SO_4$	Universal (also fat and carbon black)	No danger with exact temperature control

The powdered plant samples can also be dissolved by digesting in acids, usually HNO_3, $HClO_4$ and H_2SO_4. These acids are used either singly or in combinations of two or three acids, e.g. a di-acid combination is HNO_3 and $HClO_4$ (in 4:1 ration) or a triple acid is a mixture of HNO_3, $HClO_4$ and H_2SO_4 (in 10:4:1 ration). A triple acid combination destroys the organic matter in a shorter time without any hazard. But the method is unsatisfactory for plant materials with high Ca and in cases where S is one of the test elements. The insoluble sulphate renders the method unsuitable because of adsorption of different element ions on the precipitate and exclusion of Ca from the analysis. The use of perchloric acid in the di- or triple acid digestion mixtures results in the formation of sparingly soluble potassium perchlorate, resulting in lower estimates of K, especially when the plant material contains K, more than 1%. As such for multi element analysis, the plant materials should be digested in nitric acid alone.

4. **Interpretation of results:** The basis for plant analysis as a diagnostic technique is the relationship between nutrient concentration in the plant

and growth and production response. This relation should be significant to have complete interpretation in teem of deficient, adequate and excess nutrient concentration in the plant. Curves representing the relationship between nutrient concentration and growth response vary in shape and character depending on both the nutrient concentration in the growth medium and the plant species.

When nutrients are in deficiency range, plant growth and yield are significantly reduced and foliar deficiency symptoms appear. In this range, application of nutrient results in sharp increase in growth. In marginal range, growth or yield is reduced, but plant does not show deficiency symptoms. Sometimes the marginal range is also called transition zone. Within the marginal or transition zone lies the critical level or concentration. The critical level can be defined as that concentration at which the growth or yield begins to decline significantly.

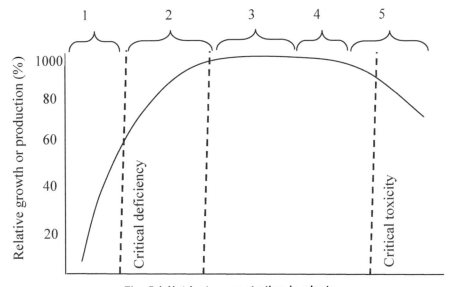

Fig. 5.1 Nutrient concentration in plants

(i) Cell sap analysis: Cell sap analysis is semi quantitative estimates of plant nutrients in field. It is more rapid than tissue analysis carried out in the laboratory. In this analysis, plant leaves or stem is chopped up and nutrients are extracted with reagents specific for each nutrient. After that, specific colour developing reagents is/are mixed with extracted plant sap. So, developed intensity of plant sap is compared with standard colour chart that indicates very low, low, medium, high and very high concentration of nutrient in plant.

Green Tissue Analysis: Three major nutrients N, P, K move from old to new tissue where deficiency symptoms are found. So plant sampling should be done from both the old and new tissue, but a test of the old tissue is sufficient. For the evaluation of the availability of plant nutrients, it is best to sample the tissue at regular intervals. If only one test is to be made, it is best to make the test when the plant is under severe nutrient stress.

Two main methods usually employed for the analysis of green plant tissues for N, P and K. **1. Paper test and 2. Glass Valve test.**

The materials and equipment required for making tests for N, P and K by the paper test and method are K-test papers (papers containing three spots of different amount of dipicrylamined; nitrate powder, a sharp knife and needle nosed pliers.

The NO_3^-N test is made by placing cut portion of green plant tissues (stem or petiole) on a clean portion of the folded test paper, adding NO_3^- powder to the tissue, and squeezing the paper and tissue together with pliers. The color of the powder turns into shades of red if NO_3^-N is present, whereas a faint pink colour indicates a low or deficient level of NO_3^-N. A complete red colour of the powder indicates a high test or sufficient nitrogen.

The phosphate-phosphorus test can be made readily by squeezing sap from freshly cut plant tissue on the paper strips. An adequate supply of P is indicated by a medium blue to dark blue colour. A light blue colour denotes P deficiency in the plant sap.

For the K tests, the plant is squeezed on the three test spots containing diplcrylamine reagent No.1 is added to given an orange colour present. The intensity in the brightness of colour determines the amount of K present. The disappearance of orange colour from all three spots on the test paper indicates a very low test, orange in the middle and bottom spots indicates a medium level and when the orange colour persists in all three spots, a high level of K is indicated.

(3) **SENSOR BASED PLANT TISSUE ANALYSIS:** Sensor based plant tissue analysis is an indirect measurement of some nutrient proportion in plant tissue. In sensor based tissue analysis, reflectance or absorption or transmission of light (having specific wavelength) by plant leaves is measured through different sensors. This technique is widely used to monitor N deficiency, plant stress and pest infestation.

Chlorophyll meter: The chlorophyll meter is a simple, portable diagnostic sensor that measures greenness or relative chlorophyll content in plant leaves.

It is based on absorbance or transmission of light (~ 650 nm) through plant leaves. Leaf chlorophyll content is highly correlated with N content in leaves, particularly over the range of yield response to applied N fertilizers.

Leaf colour chart: If chlorophyll meter is not available, simple easy to use leaf colour chart has been developed to assess N content in leaves and improve N management practices. In this method, N content in leaf is periodically assessed by comparing leaf colour with panels of critical leaf colour in chart. By this technique, farmers can manage N in field very efficiently.

Remote sensing: Remote sensing refers to the process of gathering information about an object, at a distance, without touching the object itself. Visible and near infrared sensors are commonly used to detect the plant stress related to nutrients, water and pests. When light energy strikes a leaf surface, blue and red wavelengths are absorbed by chlorophyll, while green and near infrared wavelengths are reflected by plant leaves. Reflected light is monitored by optical sensor. The contrast of light reflectance and absorption by leaves enables assessment of quantity and quality of vegetation. In general, it uses active light source to measure normalized difference vegetation index (NDVI) to determine N rate by comparing it with a nitrogen rich strip within the field. This NDVI is highly correlated with N content in crops and also other stresses caused by water and pests.

Advantages and limitations of sensor based tissue analysis:

Advantages: (i) To assess the yield limiting factors, especially N deficiency very quickly and easily under field condition thus, efficient N management practices is performed and (ii) To identify hidden hunger as well as excess application of N fertilizers, which helps in to optimize the crop yield and to reduce the pollution caused due to excess application of N fertilizers.

Limitations: (i) Initial cost is high, (ii) Skill as well as scientific knowledge is required to operate the technologies and (iii) Standardization developed for one crop or its variety may not be applicable for other crops and/or varieties due to genetic colour differences in plant.

(4) BIOLOGICAL TESTS: The biological methods consist of raising a crop or a microbial culture in a field or in a sample of the soil and estimating its fertility from the volume of crop or microbial count. If, plant growth or microbial growth is found to be satisfactorily than the soil is considered as fertile soil. Although, these methods are direct estimates of soil fertility, they are not well adopted because of lengthy (more time consuming than other methods).

Principle of biological tests: In the absence of nutrients, certain microorganism exhibits behaviour similar to that of higher plants. For example, growth of *Azotobacter* or *Aspergillus niger* refracts nutrient deficiency in the soil. The soil is rated from very deficient to not deficient in the respective elements, depending on the amount of colony growth. In comparison with methods that utilize higher plants, microbiological methods are rapid, simple and require little space. These laboratory tests are not in common use in India.

Biological methods

1. **Use of higher plants**
   a) Simple fertilizer trails on cultivator's fields
   b) Complex field experiments
   c) Methods using plants as nutrient extract ants
      i. Mitscherlich technique
      ii. Neubauer rye seedling method
   d) Use of indicator plants
   e) Visual diagnosis
   f) Foliar spray

2. **Microbiological methods**
   a) Azotobacter plaque test
   b) Aspergillus niger test
   c) Cunninghamela plaque test
   d) Carbon dioxide evaluation method

(5) SOIL TESTING: Soil testing may be defined in more restricted as well as in the broader sense. In restricted sense, *soil testing may be defined as a rapid chemical analysis to assess the different soil properties.* In a broader sense, soil testing includes five steps viz. (i) Collection of representative soil samples from the field, (ii) Preparation of samples in the laboratory (grinding, sieving, mixing etc.), (iii) Analytical procedures, (iv) Evaluation/interpretation of results and, (v) Recommendation. Soil testing provides a guideline for amendments and fertilizer needs of soils. The primary advantage of soil testing when it is compared to the plant analysis is its ability to determine the nutrients status of the soil before the crop is planted. Thus, precaution measure can be taken before growing of crop.

The following steps are involved in soil analysis

1. Sampling
2. preparation of samples

3. Analytical procedure
4. Calibration and interpretation of the results
5. Fertilizer recommendation
1. **Sampling:** Soil sampling is perhaps the most vital step for any analysis. Since, a very small fraction of the huge soil mass of a field is used for analysis; it becomes extremely important to get a truly representative soil sample from it. **"The analysis can be no better than the sample"**
2. **Preparation of sample**: Drying, grinding and sieving according to the need of analytical procedure
3. **Analytical procedure**: A suitable method is one which satisfies the following three criteria.
   i. It should be fairly rapid so that the test results can be obtained in a reasonably short period.
   ii. It should give accurate and reproducible results of a given samples with least interferences during estimation.
   iii. It should have high predictability *i.e.,* a significant relationship of test values with the crop performance.
4. **Calibration and interpretation of the results:** For the calibration of the soil test data, a group of soils ranging in soil fertility from low to high in respect of the particular nutrient are selected and the test crop is grown on these soils with varying doses of particular nutrient with basal dose of other nutrients.

The most common method is to plot soil test values against the percentage yield and to calculate the relationship between soil test values and per cent yield response

$$\text{Percent yield} = \frac{\text{Crop yield with adequate nutrient} - \text{Yield of control without addition of particular nutrient under study}}{\text{Crop yield with adequate nutrient}} \times 100$$

Major objectives of soil testing/soil analysis:
- To measure the total amount of nutrients present in the soil
- To evaluate soil fertility status for making fertilizer recommendations.
- To predict the probable crop response to applied nutrients.
- To classify soil into different types of soil groups, fertility groups for preparing soil maps and soil fertility maps.
- To identify the types and degree of soil related problems like salinity, alkalinity and acidity etc. and to suggest appropriate reclamation/ amelioration measures.

- To find out suitability for growing crops and orchard.
- To study the soil genesis.

Methods used to determine available nutrients from the soil sample:
Methods given in Table 5.1 are widely used to estimate the available nutrient status of soil (soil fertility) in most of soils. However, other specific methods may also be used depending on dominant soil properties. For example, Bray's method is used to determine available P in acid soils instead of Olsen's method.

Table 5.1: Common methods employed for soil fertility evaluation

Nutrients	Methods employed	Apparatus and/or Instruments used
SOC	Walklay and Black's titration method	Titration apparatus
Total N	Kjeldahl method	Kjeldahl's digestion cum distillation unit
Available N	Alkaline-KMnO$_4$	Kjeldahl's nitrogen distillation unit
Available P$_2$O$_5$	Olsen's method (Neutral to Alkaline soils)	Spectrophotometer
Available P$_2$O$_5$	Bray's method (Acid soils)	Spectrophotometer
Available K$_2$O	NH$_4$OAC extractant method	Flame photometer
Available S	0.15% CaCl$_2$ extractable	Spectrophotometer
Micro nutrients	DTPA extraction method	Atomic Adsorption spectrophotometer

Table 5.2: Categorization of nutrient availability as low, medium or high

SN	Nutrients	Low	Medium	High
1.	Alkaline KMnO$_4$-N (kg/ha)	<250	250-500	>500
2.	Olsens-P$_2$O$_5$ (kg/ha)	<28	28-56	>56
3.	Neutral N NH$_4$OAc-K$_2$O	<140	140-280	>280
4.	0.15% CaCl$_2$-S (mg/kg)	<10	10-20	>20
5.	DTPA extractable Fe (mg/kg)	<5	5-10	>10
6.	DTPA extractable Mn (mg/kg)	<5	5-10	>10
7.	DTPA extractable Zn (mg/kg)	<0.5	0.5-1.0	>1.0
8.	DTPA extractable Cu (mg/kg)	<0.2	0.2-0.4	>0.4
9.	Hot water soluble B (mg/kg)	<0.1	0.1-0.5	>0.5

DRIS approach: Recently Diagnosis Recommendation Integration System (DRIS) is suggested for fertilizer recommendation. In this approach, plant samples are analyzed for nutrient content and they are expressed as rations of nutrients with others. Suitable ratios of nutrients are established for higher yields from experiments and plant samples collected from farmer's fields. The nutrients whose ratios are not optimum for high yields are supplemented by top dressing. This approach is generally suitable for long duration crops, but it is being tested for short duration crops like soybean, wheat *etc.*

Objectives of plant analysis: Major objectives of plant analysis are as follows:

1. Diagnosis of nutrient deficiencies, toxicities or imbalances.
2. To know the nutrient supplying capacity of soil.
3. To calculate the amount of nutrient uptake by crop per unit area.
4. To estimate quantity of nutrients removed by crops
5. To recommend the fertilizers in a particular agro-climatic zone.
6. To monitor the response of applied fertilizers.
7. To estimate nutrient levels in the diet.

CHAPTER 6

SOIL ORGANIC MATTER (SOM)

We use the general term soil organic matter (SOM) to encompass all the organic component of soil including living organisms (the soil biomass). Soil organic matter is a complex and varied mixture of organic substances (*i.e.* partially recomposed and partially synthesized) which is derived from living entity (plants and animals). All organic substances, by definition contain the element carbon, and, on average, carbon comprises about 58 per cent of the mass of soil organic matter. In most cultivated soils, the percentage of SOM is small, but their effects on soil function are profound because, it exerts a dominant influence on many physical, chemical and biological properties of soils, especially in the surface soils. The remains of plants, animals and microorganisms are continuously broken down in the soil, and new substances are synthesized by other organisms. Over time, organic matter lost from the soil as carbon dioxide (CO_2) produced by microbial respiration. Because of such loss, repeated additions of new plant and/or animal residues are necessary to maintain soil organic matter.

6.1. SOURCES OF SOIL ORGANIC MATTER

There are mainly two sources of organic matter in the soil.

(a) *Vegetation:* Vegetation (plants) considered as a primary source of organic matter in the soil. Because large quantity of soil organic matter are originates from the plant tissues. Leaves and root of trees and crops, shrubs, grasses and other plants etc. usually supply large quantity of organic materials to the soil.

(b) *Animals*: Animals are generally considered secondary sources of soil organic matter. As they break down the original plant tissues, they contribute waste products and leave their own bodies after death.

Composition of plant residues: Plant residues are the principle material undergoing decomposition in soils and hence, are the primary source of organic matter. Green plant tissues contains from 60 to 90 % water by weight (Fig. 1). If the plant tissues are dried to remove all water, the dry matter remaining consists mostly 90 to 95 % of carbon, oxygen and hydrogen. During photosynthesis, plants obtain these elements from carbon dioxide and water. If plant dry matter is burned (oxidized), these elements become carbon dioxide and water one more. Of course, some ash and smoke will be formed upon burning, accounting for remaining 5 to 10 % of the dry matter. In ash and smoke, there may be found many nutrient elements originally taken up by the plants from the soil.

The organic compounds in the plant tissue can be grouped into broad classes. Carbohydrates, which range in complexity from single sugars and starches to cellulose, are usually the most plentiful of plant organic compounds. Lignins and poly phenols are notoriously resistant to decomposition. Certain plant parts, especially seeds and leaf coatings, contains significant amount of fats, waxes and oils which are more complex than carbohydrates but less so than lignins. Proteins contain about 16 % nitrogen and decompose easily.

6.2. NATURE OF SOIL ORGANIC MATTER

The term soil organic matter includes all the organic components of a soil. Based on these organic components, the soil organic matter is divided in to four major groups which are as follows:

a. *Living biomass*: It includes intact plant and animal tissues and microorganisms.
b. *Dead plant and animal residue*: It includes dry plant and animal residues *viz.* litter dung *etc.*
c. *Partially decomposed plant and animal residues*: It is formed by the action of soil fauna and microorganisms on the dead residue.
d. *Humus*: Humus is formed by decomposition of plant and animal residues by microorganisms. They form the largest fraction of SOM and play the dominant role in improving soil properties. Humus is defined as a complex and rather resistant (to microbial decomposition) mixture of dark-coloured amorphous and colloidal organic substances. It is derided in to two groups which are as follows:

Nonhumic substances: The terms nonhumic substances refers to the group of identifiable bio-molecules that are mainly produced by microbial action and are generally less resistant to breakdown. About 20 to 30 per cent of the humus in soils consists of nonhumic substances. For examples, proteins, carbohydrates, lignins, lipids, peptides, amino acids, alcohols *etc.* are collectively known as nonhumic substances.

Humic substances: The terms humic substances refers to the group of ill-defined, complex, resistant, polymeric, amorphous compounds and comprise 60 to 80 per cent of the SOM. They are composed of huge molecules with a molecular weight varying from 2,000 to 3, 00,000 g/mol. The three fractions of humic substances are: (i) Fulvic acid, (ii) Humic acid and (iii) Humin.

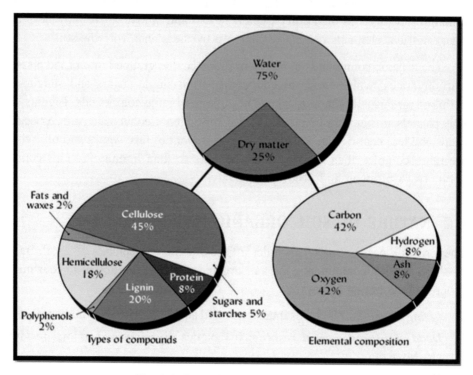

Fig. 6.1: Composition of plant residue

6.3. AMOUNT OF SOIL ORGANIC MATTER

Contents of organic matter in the soils vary and are largely dependent on the environmental conditions. In general, cultivated soils in temperate (humid) regions have high OM levels (5 to 10 %) in their surface horizon, whereas similar soils in tropics (arid and semiarid regions) have only less than 1.0 % OM levels. The reasons for these differences in OM levels have been attributed to following factors:

SOM = f **(cl, o, r, p, t...)** *i.e.* the nature and amount of soil organic matter depends on climate, organisms, relief of topography, patent materials and time.

SOIL ORGANIC MATTER (SOM)

a. **Climate**: It is not surprising that climate (temperature and rainfall) plays one of the most important roles in determining the extent of SOM. It has been observed that SOM contents decrease 2 to 3 times for every 10 °C rise in mean temperature. Apparently, the decomposition loss of SOM (to CO_2) increases more rapidly with increase in temperature up to 35 °C. Therefore, in warmer climates, SOM levels are generally lower than in cooler regions. Rainfall, an, another climatic factor which is also directly or indirectly contributes SOM by supporting vegetation and soil moisture. The rate of decomposition of SOM increases with moisture and reaches a maximum at 60 to 80 % of maximum water holding capacity.

b. **Organisms/Vegetation**: The second important factor influencing OM level in soils is the nature and amount of vegetation. Vegetation is a primary source of SOM. It provides the basic input in the form of leaf letter, branches, roots *etc.* In tropic evergreen forests, where litter input is very high, hence, the soils of such region are rich in organic matter; soils under grass cover in the same region have much lower OM level.

c. **Other factors**: Other factors which influence SOM accumulation are soil minerals and soil texture. SOM is easily lost from light textured sandy soils whereas clay soils tend to accumulate more OM. Topography influences the SOM levels mainly through soil erosion. Soils in depressions and valleys may accumulate more OM than those at slopes. If all the other factors remain constant, OM accumulations reach equilibrium levels with time.

6.4. DECOMPOSITION OF ORGANIC COMPOUNDS IN THE SOIL

Decomposition is a general term used to describe the interrelated processes by which organic matter is broken down to CO_2 and humus with a simultaneous release of nutrients. Decomposition of SOM involves the breakdown of large organic materials/ molecules in to smaller, simple components. Decomposition process takes place under both aerobic and anaerobic conditions (Fig. 2). However, the organisms involved in decomposition and its rate is discussed under separate sub heads. General decomposition process is given below:

$$\text{Complex OM (Fresh)} \xrightarrow{\textit{Microbial Decomposition}} \text{Humus} + CO_2 \uparrow \text{ and/or } CH_4 \uparrow + \text{Energy} + \text{mineral elements (Nutrients)}$$

Decomposition in aerobic Soils: When organic tissues are added in the aerobic soil, in the presence of oxygen, following three general microbial reactions takes place (Fig. 6.2):

SOIL FERTILITY AND NUTRIENT MANAGEMENT

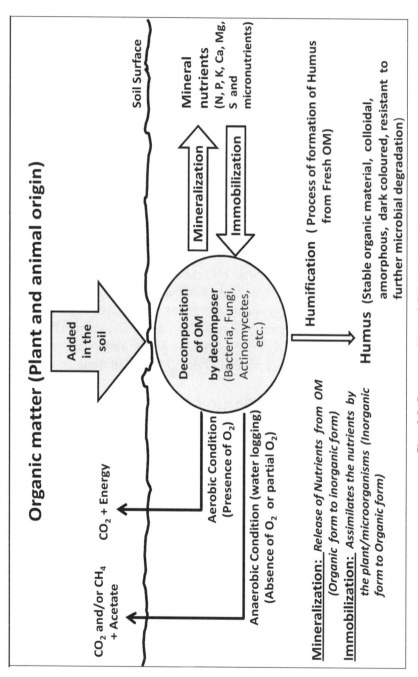

Fig.: 6.2. Decomposition of OM in the soil

a. **Enzymatic oxidation:** Enzymatic oxidation is the process in which carbon compounds of added organic materials is converted into carbon dioxide, water, and energy due to enzymatic activities in the presence of oxygen.

b. **Mineralization/Immobilization:** Release (mineralization) and/or immobilization of essential nutrient elements, such as nitrogen, phosphorus, sulphur etc by a series of specific reactions.

c. **Formation of humus:** Formation of organic compounds very resistant to microbial degradation either through modification of compounds in the original tissue or by microbial synthesis.

In aerobic decomposition process, many intermediate steps are involved. When plant protein decay, they not yield carbon dioxide and water only, but amino acids and other organic acids are also released in the soils which stimulate plant growth. The process that release elements from organic compounds to produce inorganic (mineral) forms is known as mineralization, usually last step in the overall decomposition process which is important source of essential available nutrients in the soils. If the availability of mineral nutrient, especially nitrogen, is low in organic tissues for decomposers (microorganisms) than their requirement, they utilize mineral elements from the soil for their body building and energy and converted them in to organic form; the process is called as immobilization.

Decomposition in Anaerobic Soils: Without sufficient oxygen present in soils, aerobic organisms cannot function, so anaerobic or facultative organisms become dominant. Under low-oxygen or anaerobic conditions, decomposition takes place much more slowly than when oxygen plentiful. Hence, wet, anaerobic soils tend to accumulate large amount of organic matter in a partially decomposed condition.

Mineralization	Immobilization
• It is the process in which organic form of elements is converted in to inorganic (mineral) form is called mineralization.	• It is the process in which inorganic form of element is converted in to organic form is called immobilization.
• Release of essential nutrients in the soils is an example of mineralization	• Immobilize the nutrients from the soil by organisms is an example of immobilization
• It increase soil fertility	• It decrease the soil fertility
• If, the C:N ratio of added organic matter is less than 20:1, then net mineralization takes place in soil.	• If, the C:N ratio of added organic matter is greater than 30:1, then net immobilization takes place in soil.

The products of anaerobic decomposition include a wide variety of partially oxidized organic compounds. Anaerobic decomposition releases relatively little energy for the organisms involved, therefore the end products *i.e.* alcohol and methane still contain much energy. Some of the products of anaerobic decomposition are of concern because they produce foul odors or inhibit plant growth. The methane gas produced during anaerobic condition (by various methanogenic bacteria and archaea) is major contributor to the greenhouse effect.

Factors affecting rate of SOM decomposition/mineralization

When any organic residues/organic materials are added in the soil, they are subjected to microbial decomposition. The time required to complete decomposition process may range from days to years, depending mainly on two broad factors: (i) Environmental conditions and (ii) quality of added organic materials in the soils.

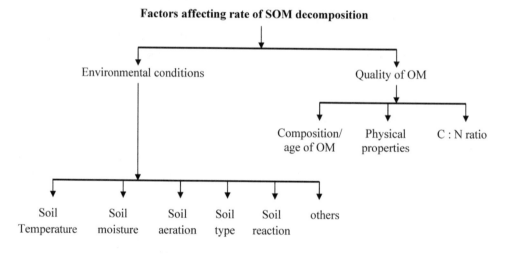

Environmental conditions in the soils: The soil environmental conditions viz. soil temperature, soil moisture, soil aeration and available nutrients plays an important role in OM decomposition.

- **(a) *Temperature:*** Extreme low or high temperature retards rate of decomposition. The optimum temperature range for maximum speed of decomposition of OM is 25 to 35 °C

- **(b) *Soil moisture:*** Both dry soil and water saturated soils (anaerobic conditions) reduce plant growth and microbial decomposition. Near or slightly wetter than field capacity moisture conditions (60 to 80 % of maximum water holding capacity moisture content) are most favorable for OM decomposition.

SOIL ORGANIC MATTER (SOM)

(c) Soil aeration: Aerobic decomposer also requires oxygen for their activity. Hence, sufficient soil aeration is also essential for rapid decomposition of SOM.

(d) Soil type: Soil texture indirectly influences the rate of OM decomposition. Soil organic matter tends to increase as the clay content increases. This increase depends on two mechanisms. First, bonds between the surface of clay particles and organic matter retard the decomposition process. Second, soils with higher clay content increase the potential for aggregate formation.

(e) Soil reaction: Soil reaction influences the activity of soil microorganisms and hence influences the decomposition of organic residues. Most of the microbes grow best at pH 6.0 to 8.0, but are severely inhibited below pH 4.5 and above pH 8.5.

(f) Others: Lack of nutrients particularly nitrogen (N) slows decomposition rate. Some toxic elements *viz.* Al, B, Mn, *etc.* and excess amount of salts in soils also retards the rate of OM decomposition processes.

Quality of organic residues: Like environmental factors, the quality of organic residues added to the soil *viz.* organic components, physical factors and C: N ratio of organic materials, are also play an important role in microbial decomposition and subsequent release of nutrients in the soil.

(a) Composition of organic matter: Different plant and animal organic residues contain different group of organic compounds as given below. Organic compounds can be listed in terms of ease of decomposition as follows:

	Composition of organic matter	Ease of decomposition
1.	Sugars, Starches and Simple Proteins	Rapid decomposition
2.	Crude protein	↑
3.	Hemi cellulose	
4.	Cellulose	
5.	Fats and Waxes	↓
6.	Lignins and phenolic compounds	Very slow decomposition

It is evident that different constituents of organic residues decompose at different rates. Simple sugars, starches, amino acids, most proteins decomposed very within a short periods (months). Whereas, lignins and phenolic compounds are notoriously resistant to microbial decomposition and may take years for completion of decomposition. For example, legume crop residues are decomposed easily than cereals/grasses/woods. Because legume crop residues

84 SOIL FERTILITY AND NUTRIENT MANAGEMENT

contain more amount of protein component and hence decompose very easily than cereals/grasses/woods.

(b) Physical factors: The location and particle size of organic residues are important physical factors that influence the rate of decomposition. When residues are incorporated in to the soil by means of organisms and/or tillage operation are decomposed rapidly than surface placement as like mulch. Residue particle size is another physical factor that also affects the decomposition process. The smaller sized particles are decomposed more rapidly than large sized particles because of more surface area available for microbial degradation.

(c) C : N ratio of organic materials: When plant residues or other organic materials (FYM, Compost *etc.*) added to the soil, the differences in C/N ratio have pronounced effect on the rate of decomposition. Decomposition of lower C/N ratio materials such as fresh grasses, legumes, and manure tends to be more rapid than high C/N ratio material. Nitrogen will not be a limiting factor and these materials also contain more easily decomposed compounds such as sugars, amino acids and proteins.

6.5 C : N RATIO

The C : N ratio or C/N ratio (carbon-to-nitrogen ratio) is a ratio of the mass of carbon to the mass of nitrogen in a substance (OM, organisms, soils *etc.*). The average carbon content of plant residues is around 42 % (Fig.:1). The nitrogen content of plant residues is much lower and varies widely (from < 1.0 to > 6.0 %). The C/N ratio in plant residues ranges from between 10:1 to 30:1 in legumes and young green leaves to as high as 600:1 in some kind of Sawdust. Whereas, the C/N ratio of soil microorganisms are ranges from 5:1 to 11:1 with an average of 8:1. While, in cultivated soils, generally, the C/N ratio is stabilized at 10:1 to 12:1 (Average: 11:1) after decomposition of added organic matter. When plant residues or other OM (FYM, compost *etc.*) added to the soil, these differences in C/N ratio have pronounced effect on the rate of decomposition and also results in the release (mineralization) or immobilize (immobilization) of soil available nutrients especially nitrogen.

Calculation of C : N ratio: The C/N ratio of organic material is calculated as follows:

If the green gram residue contains 50 % C and 2 % N then,

$$C:N \text{ ratio} = \frac{\text{Mass of Carbon}}{\text{Mass of Nitrogen}} = \frac{50}{2} = 25$$

Hence, C:N ratio of green gram residue is 25 : 1

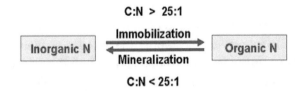

C:N Ratio of Organic Material Affects Rate of Activity

Residue with High C:N Ratio -	Residue with Low C:N Ratio -
Straw, Cornstalks	Alfalfa, Soil Organic Matter
Available nitrogen is tied up	Available nitrogen is released

Fig. 6.3: When organic material (FYM, Compost etc.) having different C:N ratio are added in the soils they influences the nutrient availability especially nitrogen. Wider C:N ratio (> 25:1) immobilize the available soil nitrogen and cause nitrate depression period (decreases available N in soil for some time). Whereas, lower C:N ratio mineralize the N and increases nitrogen availability in the soil.

Explanation: Degradation of organic material involves in important balance between carbon and nitrogen in the material being degraded (OM), in the degraders (Soil microbes), and in the soil. Soil microbes, like other organisms, require balance nutrients from which to build up their cells and extract energy. Soil organisms need carbon for building essential compounds and to obtain energy. However, organisms must also obtain sufficient nitrogen to synthesize nitrogen-containing cellular components, such as amino acids, enzymes and DNA. On the average, soil microbes must incorporate in to their cells about eight parts of carbon for every one part of nitrogen. Because only about one-third of carbon metabolized by microbes is incorporated in to their cells (the remainder is respired and lost as CO_2), the microbes need to find about 1 g of nitrogen for every 24 g of carbon in their "food".

Example: Bacteria and fungi have an average C/N ratio in their cells of about 8:1 (8 g C and 1 g N). This ratio must be maintained. If fresh organic material has a C/N ratio of around 24:1 (24 g C and 1 g N), this provides exactly the ratio needed to keep the bacteria and fungi C/N ratio at 8:1. This is because with 2/3 of the carbon (16 g C) being lost as carbon dioxide, the C/N ratio of what the microbes actually use is very close to 8:1.

Importance/practical significance of C/N ratio in cultivated soils

When organic manures (FYM, cakes, biocompost *etc.*) are added in the soil, they start to decay by soil microorganisms under favourable conditions (as above mentioned). The rate of organic material decay and nitrogen availability to growing plant is greatly influenced by C/N ratio of organic material which is being added in to soil as organic manures. In other words, the nitrogen availability is controlled by two important processes (mineralization and immobilization) operating in the soils when organic manures having different C:N ratio, which are as follows:

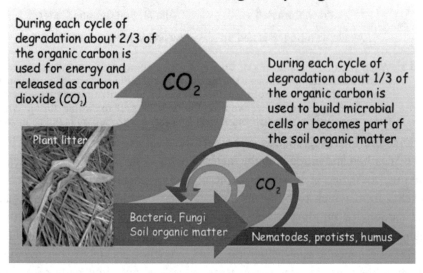

Fig. 6.4: Figure shoving the degradation of organic material involves in important balance between carbon and nitrogen in the material being degraded, in the degraders, and in the soil. When fresh litter is degraded, about 2/3 of the carbon is released as carbon dioxide, and about 1/3 goes into building new biomass. This cycle repeats over and over until the material is degraded to stable soil humus.

Immobilization (C : N ratio greater than 30 : 1): When organic materials containing C/N ratio greater than 30:1 are added to the soils as manure, the immobilization will results. It means plant available form of inorganic nitrogen is converted in to organic forms and therefore available N content in soil is decreased (Fig. 5). (Because of more demand of N for decomposing microorganisms, so, they immobilize soil available N for balancing their C/N ratio of 8:1. Therefore, available N content in soil is decreased)

This decreased available N period is called *"Nitrate Depression Period"*. This period may last for few weeks or few months depending upon nature of organic material and environmental conditions. Nitrate depression period adversely affects the germination of seeds, seedling and plant growth. Plants will remain stunted, chlorotic, and growth is drastically reduced.

How to overcome nitrate depression period?

1. Delay the sowing/planting after application of high C/N ratio organic material in the soil.
2. Provide optimal environment to decompose organic residue rapidly.
3. In order to lower down C/N ratio, compost the high C/N ratio organic material by different composting processes before incorporating in the soil.
4. Required quantity of Nitrogen should be applied from outer source (Fertilizers) to meet the demand of raising crops and degraders.

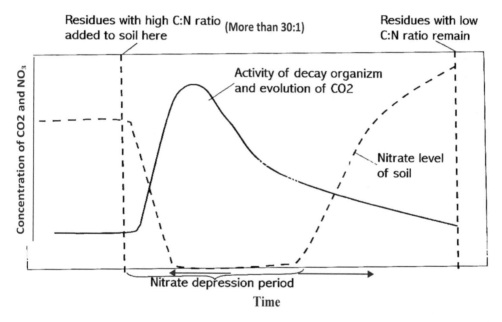

Fig. 6.5.: Relationship showing the time of organic matter decomposition and concentration of release of CO_2 by organisms (microbial respiration) and NO_3^- in the soil.

Mineralization (C:N ratio less than 20:1): When the organic materials of low C/N ratio are added in the soil as organic manure, the mineralization will results, it means that organic form of N present in manure is converted in to inorganic N and therefore, available N content in soil is increased.

Benefits: Addition of organic manure containing Low C:N ratio, within a few days, the available N content in soil increased, which results in increases in plant growth

Neither Mineralization nor Immobilization (C:N ratio between 20:1 to 30:1): When the organic materials of C:N ratio in-between 20:1 to 30:1 are added in the soil, both the processes will operates equally. At initial stage (few weeks), the microorganisms not utilize the soil available N nor release the N in the soil. But letter on, soil available N is increases due to net mineralization.

Some important biochemical processes operating in the soil during the decomposition of organic matter OR Nitrogen transformation and translocation:

Nutrients (especially N) present in soil or added to the soil (as a manures or fertilizers), are subject to several changes (transformations) that decides their availability to plant and movement in the soil and in the environment. Understanding these transformation/translocation processes will help you make the best use of manure and fertilizers to meet crop needs while safeguarding the environment. In general, the N cycle processes of fixation, mineralization, ammonification and nitrification increase plant available N. While, denitrification, volatilization, immobilization, and leaching result in permanent or temporary N losses from the root zone. Read on for specifics about each of the N cycle/transformation processes (Fig. 6.3).

Transformation: It is the process of conversion of one form of nutrients into another form

Mineralization: Broadly, mineralization is the process by which organic form of nutrients are converted in to inorganic form. In this process, the nutrients especially N and S present in complex organic materials are converted in to mineral (inorganic) form through the microbial decomposition. More specifically, for individual nutrients mineralization, we used the term mineralization with notation of nutrient symbol or name *viz.* Nitrogen mineralization or N mineralization and so on for other nutrients also. N mineralization process includes two steps *viz.* Aminization and ammonification.

Aminization: It is the process in which complex organic nitrogenous (proteins) compounds are converted in to in to simple N compounds (amino acids and amines). It is a process of enzymatic digestion by bacteria, actinomycetes and fungi.

Ammonification: Ammonification is the process by which simple N compounds (amino acids and amines) are converted in to ammonia (NH_3) and then ammonium N (NH_4^+ - N). NH_4^+ ion produced through ammonification is subject to several fates as follow:

- Converted in to NO_2^- and NO_3^- (*Nitrification*)
- Absorbed directly higher plants (*N uptake*)
- Utilized by soil microorganisms (*Immobilization*)

- Converted back in to NH_3 and release to atmosphere under alkaline condition (*Volatilization*)
- Fixed by some silicate clay minerals (NH_4^+ *fixation*)

All the organisms (A diverse population of aerobic and anaerobic bacteria, fungi and actinomycetes including soil fauna) present in soil are directly or indirectly involved in decomposition/mineralization of organic materials. However, heterotrophic microorganisms (that require organic C for energy) are dominantly involved in this complex process under neutral and alkaline soils while, fungi are dominant in acid soils. Mineralization operates under both the aerobic and anaerobic conditions but it is much faster in aerobic condition (presence of oxygen) than the anaerobic soil environments.

Immobilization: Immobilization is the reverse of mineralization. It refers to the process in which inorganic form of nutrients are converted in to organic forms through soil microorganisms. In N Immobilization, inorganic N compounds (NH_4^+ and NO_3^-) are converted to organic N by soil organisms and therefore become unavailable to crops.

- Incorporation of organic materials with a high carbon to nitrogen ratio (e.g. sawdust, straw, *etc.*), will increase biological activity and cause a greater demand for N, and thus result in N immobilization.
- Immobilization only temporarily reduces the N availability. When the microorganisms die, the organic N contained in their cells is back converted in to available form through the process of *mineralization* and *nitrification*.

Nitrification: Nitrification is the process of enzymatic oxidation in which ammonium N (NH_4^+ - N) is transformed in to Nitrate N (NO_3^- - N) by a specialized group of bacteria. This process requires supply of ammonium ion and oxygen to make nitrite and nitrate ions and is therefore favoured well drained aerobic soil condition. The bacteria involved in nitrification process are classified as *autotrophic* bacteria (Autotrophs) because they obtain their energy from oxidizing ammonium ions rather than organic matter. Nitrification consists of two sequential steps: (1) the first step results in conversion of ammonium (NH_4^+) to nitrite (NO_2^-) by group of *Nitrosomonas* bacteria. (2) In second step, the nitrite (NO_2^-) so formed is immediately converted in to nitrate (NO_3^-) by group of *Nitrobacter* bacteria. The second step of nitrification process is fast than first one therefore; very little nitrites accumulate in soil. This is fortunate, because even at low concentration, nitrite is quite toxic to most plants.

So formed nitrate ion (NO_3^-) through nitrification process is subject to several fates which are as follows:

90 SOIL FERTILITY AND NUTRIENT MANAGEMENT

- Absorbed directly higher plants (*N uptake*)
- Utilized by soil microorganisms (*Immobilization*)
- Loss in atmosphere as NO, N_2O and N_2 under anaerobic soil (*Denitrification*)
- Loss with running water (*Runoff losses*)
- Loss in atmosphere as NH^3 under alkaline condition (*Volatilization*)

Losses of N from the root zone: There are various ways by which N is losses from the root zone of soil which are as follows:

1. *Gaseous losses of nitrogen:* Under certain some abnormal conditions of soil, inorganic N can be converted into gases primarily by denitrification and volatilization and lost to the atmosphere.

   a. **Denitrification:** Nitrification is the reverse of nitrification and operates under anaerobic condition (water logged soil). In this process, the nitrate N is converted in to gaseous form of N (NO, N_2O and N_2) and lost to atmosphere by a series of biochemical reduction reactions through the anaerobic denitrifying bacteria. Large populations of denitrifying bacteria present in soils are carrying out this process and are mostly facultative anaerobic heterotrophic bacteria (*Pseudomonas, Bacillus and Paracoccus*) while, *Thiobacillus denitrificans* and *Thiobacillus thioparus* are autotrophic bacteria.

   b. **Volatilization:** Ammonia gas (NH_3) can be produced in soil is lost to the atmosphere is termed as volatilization loses of N. Ammonia is released in soil as follows: (a) during the decomposition of SOM produce NH_3 (b) Under alkaline condition, ammonical ion is converted to ammonia and (c) Application of N fertilizers especially urea and anhydrous ammonia. The volatilization losses increase at higher soil pH and conditions that favor evaporation (e.g. hot and windy).

   Reason: 25 % more recommended dose nitrogenous fertilizers are advised in alkaline soil.

2. *Leaching losses of N:* Loss of dissolved N compounds with percolating water from surface soil to below the root zone is termed as leaching losses of N. Of the total N leaching loss, more than 90 per cent N is lost in the form of NO_3^-, because firstly, it is very soluble in water and highly mobile in soil and secondly, NO_3^- is negatively charged ions and therefore, it is not adsorbed on negatively charged colloidal particles and free to move with water (AEC of normal soil is much less than CEC). In contrast, NH_4^+ is being a cation, gets adsorbed on clay soil colloids and therefore, its loss due to leaching is much less. A large amount of amide (NH_2) form of N also leached when urea fertilizers is applied following heavy irrigation or rainfall. Leaching losses of N is more pronounced in light texture soils than heavy textured soils.

3. **Runoff losses of N:** Loss of nitrogen with runoff water moving across the soil surface is called as runoff losses of N. Mostly, dissolved N compounds are lost by this way. The N lost by leaching losses and/or runoff losses not only results in decrease in soil fertility but also pollute the ground/drinking water and ultimately our environment.

4. **Crop removal of N**: Nitrogen taken up by the plant cannot be considered as its losses in real sense as the absorbed N is converted in to proteins and amino acids in plant which is utilized by mankind and animals as food and fodder. Nevertheless, soil losses that much N from the field in the form of harvested economic yield.

   NH_4^+ Fixation: Because of ionic diameter of ammonium ion (2.96 angstrom) is closer to some interlayer of 2:1 type silicate minerals especially vermiculites and mica, a part of NH_4^+ ions present in soil or applied as a fertilizers is fixed in the interlayer of these silicate clay minerals is called ammonium fixation by clay minerals (K^+ ion is also fixed in same way by vermiculites and mica silicate clay minerals). These fixed ammonium ions are relatively unavailable to plant however, it released slowly by other ions (like K^+) and/or utilized by nitrifying bacteria and become available to plant. Therefore, NH_4^+ fixation mechanism may be considered an advantage for crop production point of view as it prevents the leaching, volatilization, denitrification and runoff losses of N.

Organisms involved in organic matter decomposition

All the organisms present in soils are involved in organic material decomposition. Some important organisms (Fauna and Flora) and their major role are given below.

Sr. No.	Organisms	Major role in organic matter decomposition
1.	Earthworms	• Mix fresh organic materials into the soil and brings organic matter into contact with soil microorganisms
2.	Soil insects and other arthropods	• Shred (cut up) fresh organic material into much smaller particles
		• Allows soil microbes to access all parts of the organic residue
3.	Bacteria	• Population increases rapidly when organic matter is added to soil
		• Quickly degrade simple compounds *viz.* sugars, proteins and amino acids

[Table Contd.

92 SOIL FERTILITY AND NUTRIENT MANAGEMENT

Contd. Table]

Sr. No.	Organisms	Major role in organic matter decomposition
4.	Fungi	• Have a harder time degrading cellulose, lignin, starch • Cannot get at easily degradable molecules that are protected • Grow more slowly and efficiently than bacteria when organic matter is added to soil • Able to degrade more complex organic molecules such as hemicellulose, starch and cellulose. • Give other soil microorganisms access to simpler molecules that were protected by cellulose or other complex compounds.
5.	Actinomycetes	• The cleanup crew • Become dominant in the final stages of decomposition • Attack the highly complex and decay resistant compounds viz. Cellulose, Chitin (insect shells), Lignin and Waxes
6.	**Predators:** Protists and Nematodes	• Feed on the primary decomposers (bacteria, fungi, actinomycetes) • Release nutrients (nitrogen) contained in the bodies of the primary decomposers

6.6. ROLE OF ORGANIC MATTER IN THE SOIL

Addition of organic matter improves the most of physical, chemical and biological property of soils which results in improves the soil health and crop productivity.

1. Organic matter binds soil particles into structural units called aggregates. These granular aggregates increases total porosity of soils and make soil easier to cultivate.

2. Increases in granular type soil aggregates provides congenial environment for crop growth by balancing the proportion of macro and micro porosity of soils.

3. In sandy soils, addition of organic matter increases the micro porosity of soils and therefore increases the water holding capacity of sandy soils. While in case of clay soils, it increases the macro porosity which facilitates more aeration and root growth by improving the drainage capacity of clay soils.

4. Addition of organic matter also improves the infiltration rate, permeability, hydraulic conductivity and water stable aggregates (WSA) of soil.

5. Addition of organic matter in soil or surface mulching with coarse organic material reduces the losses of soil and nutrients by erosion. Because of formed aggregates are resistant to erode and increases the infiltration rate of which percolate more water in to the soil.

SOIL ORGANIC MATTER (SOM)

6. Surface mulching with coarse organic matter, lower soil temperatures in the summer and keep the soil warmer in winter.
7. The organic matter serves as a source of energy for the growth of soil microorganisms.
8. Organic matter act as a store house of essential nutrients. Upon decomposition, organic matter supplies the nutrients needed by growing plants, as well as many hormones and antibiotics.
9. Fresh organic matter has a special function in making soil phosphorus more readily available in acid soils.
10. Organic acids released from decomposing organic matter help to reduce alkalinity of soils.
11. Fresh organic matter supplies food for such soil life as earthworms, ant and rodents. These microorganisms improve drainage and aeration status of soil.
12. During the decomposition of organic material, some organic acids and carbon dioxide are released in soil which helps to dissolve unavailable nutrients and make them more available to growing plants.
13. Humus (highly decomposed organic matter) provides a store house for the exchangeable cations viz. NH_4, K, Ca, Mg and micronutrient cations and prevented them from leaching
14. Organic matter acts as a buffering agent. Buffering checks rapid chemical changes in pH.

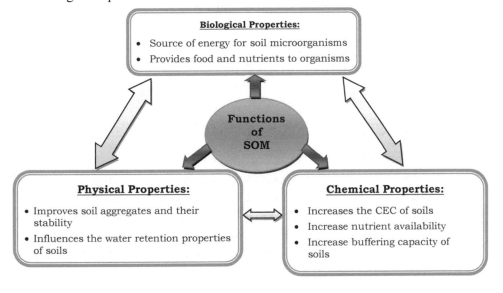

CHAPTER 7

ORGANIC MANURES

The word manure derived from the French '*Manoeuvrer*', means to manipulate, to work, to produce crop. In general manure means excreta of animals. The term bulky organic manure generally includes those materials of natural origin, organic composition having greater volume per unit content of nutrients and being used to increase the nutrient status of the soils as well as organic matter content of soils. They are obtained mainly as natural products.

"The manures are organic in nature, plant or animal origin and contain organic matter in large proportion and plant nutrients in small quantities and used to improve soil productivity by correcting soil physical, chemical and biological properties."

The materials included in this group are farmyard manure, compost, sewage sludge and green manure. Of these FYM, compost and green manure are the most important and widely used bulky organic manures.

7.1. CHARACTERISTICS OF MANURES

Manure required in large quantity bulky and costly. Nutrients are slowly available upon decomposition. It has long lasting effect on soil and crop. No salt and adverse effect. Manure is organic matter used as organic fertilizer in agriculture. Manures contribute to the fertility of the soil by adding organic matter and nutrients, such as nitrogen, that are trapped by bacteria in the soil. Higher organisms then feed on the fungi and bacteria in a chain of life that comprises the soil food web.

7.2. CLASSIFICATION OF ORGANIC MANURES

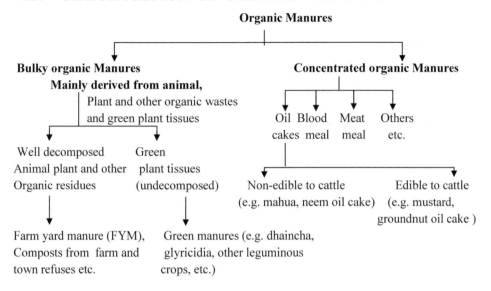

7.2.1. Bulky Organic Manures

The manures which supply plant nutrients in small quantities, and organic matter in large quantities as compared to concentrated organic manures. eg. FYM, compost, nigh soil, green manures.

Effect of bulky organic manures on soil properties

The effect of bulky organic manures on soil is threefold:

1. Bulky organic manures increase organic matter content and hence improve the physical properties of soil. This effect is very important in case of most of our arable land. Such manures increase the humus content of soil at least temporarily and consequently the water holding capacity of sandy soils is increase and drainage of clayey soils is improved.

2. Since these manures contain plant nutrients, they have a direct effect on plant growth, like any other commercial fertilizers. Bulky organic manures contain plant nutrients in small quantities, therefore large quantities of them need to be applied per hectare. Besides the major nutrients, bulky organic manures also contain traces of micronutrients.

3. Bulky organic manures provide food for soil micro-organisms. This increases activity of microbes which in turn help to convert unavailable plant nutrients into available forms.

A. Farm Yard Manure (FYM)

The FYM refers to the decomposed mixture of dung and urine from farm animals, mainly sheep, cattle and poultry. This is one of the oldest manure known and is highly valued for its many of the beneficial properties that are said to be produced when this manure is added to the soil. It not only adds the constituents to the soil but also adds organic matter to the soil.

On an average well rotted FYM contains 0.5% N_s 0.2% P_2O_5 and 0.5% K_2O.

FYM is one of the most important agricultural by products. Unfortunately, however nearly 50 per cent of the cattle dung production in India today is utilized as fuel and is thus lost to agriculture.

Average percentage of N, P_2O_5 and K_2O in the fresh excreta of farm animals:

Excreta of		N (%)	P_2O_5 (%)	K_2O (%)
Cows and bullocks	Dung	0.40	0.20	0.10
	Urine	1.00	Traces	1.35
Sheep and goat	Dung	0.75	0.50	045
	Urine	1.35	0.05	2.10
Buffalo	Dung	0.26	0.18	0.17
	Urine	0.62	Traces	1.61
Poultry	–	1.46	1.17	0.62

Among bulky organic manures, poultry manure generally contains more amounts of nutrients as compared to others. Moreover, urine portion of all farm animals contains more amounts of N and K as compared to the dung portion.

Factors Affecting Nutritional Build up of FYM

The following factors affect the composition of FYM

1. **Age of animal:** Growing animals and cows producing milk retain in their system nitrogen and phosphorus required for productive purposes like making growth and producing milk and the excreta do not contain all the ingredients of plant food given in the feed. Old animals on the downgrade waste their body tissues and excrete more than what they do ingest.
2. **Feed:** When the feed is rich in plant food ingredients, the excreta produced is correspondingly enriched.
3. **Nature of Litter Used:** Cereal straw and leguminous plant refuse used as litter enriched the manure with nitrogen.

4. **Ageing of Manure:** The manure gets richer and less bulky with ageing.
5. **Manner of Making and Storage:** In making and storage losses are in various ways. (see 'Losses in FYM).

Losses during handling and storage of FYM:

(I) Losses during handling: FYM consists of two original components the solid or dung and liquid or urine. Both the components contain N, P_2O_5 and K_2O the distribution of these nutrients in the dung and urine is shown in figure below:

Approximately half of N and K_2O is in the dung and the other half in urine. By contrast, nearly all of the P_2O_5 (96%) is in the solid portion. To conserve N, P_2O_5 and K_2O, it is most essential that both the parts of cattle manure are properly handled and stored.

i) **Loss of liquid portion or urine:** Under Indian conditions the floor of the cattle shed is usually un-cemented or Kachha. As such the urine passed by animals during night gets soaked into the Kachha floor. When the animals, particularly bullocks, are kept in the fields during the summer season, urine gets soaked into soil. But during remaining period cattle are kept in a covered shed and therefore the Kachha floor soaks the urine every day. Large quantities of nitrogen are thus lost through the formation of gaseous NH_3. The following reactions take place:

$$NH_2 CO NH_2 + 2H_2O \longrightarrow (NH_4)_2CO_3$$
Urea in urine → Ammonium carbonate

$$(NH_4)_2 CO_3 + 2H_2O \longrightarrow 2NH_4OH + H_2 CO_3$$

$$NH_4OH \longrightarrow NH_3 \uparrow + H_2O$$

Gaseous Ammonia

The smell of NH_3 in the cattle shed clearly indicates the loss of N.

No special efforts are made in India to collect the liquid portion of the manure.

ii) **Loss of solid portion or dung:** It is often said that 2/3 of the manure is either utilized for making cakes or is lost during grazing, the remaining manure is applied to the soil after collecting in heaps. Firstly, the most serious loss of dung is through cakes for burning or for use as fuel- Secondly, when milch animals go out for grazing, no efforts are made to collect the dung dropped by them, nor is this practicable, unless all milch animals are allowed to graze only in enclosed small size pastures.

SOIL FERTILITY AND NUTRIENT MANAGEMENT

(II) Loss during storage: Mostly, cattle dung and waste from fodder are collected daily in the morning by the cultivators and put in manure heaps in an open space outside the village. The manure remains exposed to the sun and rain. During such type of storage, nutrients are lost in the following ways:

i) **By leaching:** Losses by leaching will vary with the intensity of rainfall and the slope of land on which manure is heaped. About half of portion of N and P_2O_5 of FYM and nearly 90% of K are water soluble. These water soluble nutrients are liable to get washed off by rain water.

ii) **By Volatilization:** During storage considerable amount of NH_3 is produced in the manure heap from

   i) the decomposition of urea and other nitrogenous compounds of the urine and

   ii) the much slower decomposition of the nitrogenous organic compounds of the dung. As the rotting proceeds, more and more quantity of ammonia is formed. This NH_3 combines with carbonic acid to form ammonium carbonate and bicarbonate. These ammonium compounds are unstable and gaseous NH_3 may be liberated as indicated below:

   1. Urea and other nitrogenous compounds in urine and dung $\xrightarrow{\text{microbial decomposition}}$ $NH_3 \uparrow$
   2. $2NH_3 + H_2CO_3 \longrightarrow (NH_4)_2 CO_3$
   3. $(NH_4)_2 CO_3 + 2H_2O \longrightarrow 2NH_4OH + H_2CO_3$
   4. $NH_4OH \longrightarrow NH_3 + H_2O \uparrow$

Loss of NH_3 increases with

i) the increase in the concentration of ammonium carbonate

ii) increase in the temperature and

iii) air movement

Improved Methods of Handling FYM

It is practically impossible to check completely the losses of plant nutrients and organic matter during handling and storage of FYM. However, improved methods could be adopted to reduce such losses considerably.

Among these methods are described here under:

i) Trench method of preparing FYM

ii) Use of gobar gas-compost plant

iii) Proper field management of FYM

iv) Use of chemical preservatives

i) **Trench method of preparing FYM:** This method has been recommended by Dr. C. N. Acharya. The manure preparation should be carried out in trenches, 20 to 25 ft. long, 5 to 6 ft. broad and 3 to 3.5 ft. deep. Cattle shed and portions of litter mixed with earth if available. When trench is completely filled up, say in about three months time.

ii) **Use of gobar gas compost plant:** Methane gas is generated due to anaerobic fermentation of the most common organic materials such as cattle dung, grass, vegetable waste and human excreta. Gobar gas and manure both are useful on farms as well as in homes. A few advantages of this method are giving below:

1) The methane gas generated can be used for heating, lighting and motive power.

2) The methane gas can be used for running oil engines and generators

3) The manure which comes out from the plant after decomposition is quite rich in nutrients. N 1.5%, P_2O_5- 0.5%, K_2O- 2.0%

4) Gobar gas manure is extremely cheap and is made by locally available materials.

Superiority of gobar gas compost plant over traditional method

1000 Kg fresh dung manure obtained by

Sr. No.	Particulars	Traditional method	Gobar gas plant
1.	Loss of OM	500 Kg	270 Kg
2.	Loss of N	1.25 Kg	Nil
3.	Final manure	500 Kg	730 Kg
4.	% N	0.5%	1.5%
5.	Additional advantage	–	2000 C.ft. gas for cooking

iii) **Proper field management of FYM:** Under field conditions, most of the cultivators unload FYM in small piles in the field before spreading. The manure is left in piles for a month or more before it is spread. Plant nutrients are lost through heating and drying. To derive maximum benefit from FYM, it is most essential that it should not be kept in small piles in the field before spreading, but it should be spread evenly and mixed with the soil immediately.

iv) **Use of Chemical Preservatives:** Chemical preservatives are added to the FYM to decrease N losses. To be most effective, the preservatives are applied in the cattle yard to permit direct contact with the liquid portion of excreta or urine. This has to be done because the loss of N from urine starts immediately. The commonly used chemical preservatives are I) Gypsum and ii) Super phosphate.

The value of gypsum in preserving the N of manure has been known and it has been used for many years in foreign countries. The reaction of gypsum with ammonium carbonate (intermediate product from decomposition of urea present in urine) is :

$$(NH_4)_2 CO_3 + CaSO_4 \longrightarrow CaCO_3 + (NH_4)_2 SO_4$$

As long as the manure is moist, no loss of NH_3 will occur, but if the manure becomes dry, the chemical reaction is reversed and the loss of NH_3 may occur. As such, under Indian conditions, use of gypsum to decrease N losses, does not offer a practical solution.

Superphosphate has been extensively used as a manure preservative:

$$2CaSO_4 + Ca(H_2PO_4)_2 + 2(NH_4)_2 CO_3 \longrightarrow Ca_3(PO_4)_2 + 2(NH_4)_2 SO_4 + 2H_2O + 2CO_2$$

In this reaction, tricalcium phosphate is formed which does not react with ammonium sulphate, when manure becomes dry. As such, there is no loss of NH_3.

Since FYM becomes dry due to high temperature under Indian conditions, the use of superphosphate will be safely recommended as a preservative to decrease N losses.

Use of superphosphate as a chemical preservative will have three advantages

1. It will reduce loss of N as ammonium from FYM.
2. It will increase the percentage of P in manure thus making it a balanced one.
3. Since, tricalcium phosphate produced with the application of superphosphate to the FYM is in inorganic form, which is readily available to the plants; it will increase the efficiency of phosphorus.

It is recommended that one or two pounds of SSP should be applied per day per animal in the cattle shed where animal pass urine.

Supply of plant nutrients through FYM

On an average, FYM applied to various crops by the cultivators contains the following nutrients:

% N : 0.5 % P_2O_5 : 0.2 % K_2O : 0.5

Based on this analysis, an average dressing of 10 tones of FYM supplies about

50 Kg N
20 Kg P_2O_5
50 Kg K_2O

ORGANIC MANURES 101

All of these quantities are not available to crops in the year of application, particularly N which is very slow acting. Only 1/3 of the N is likely to be useful to crops in the first year. About 2/3 of the phosphate may be effective and most of the potash will be available. This effect of FYM application on the yield of first crop is known as the direct effect of application. The remaining amount of plant food becomes available to the second, third and to a small extent to the fourth crop raised on the same piece of land. This phenomenon is known as the residual effect of FYM.

When FYM is applied every year, the crop yield goes on increasing due to direct plus residual effect on every succeeding crop. The beneficial effect is also known as cumulative effect.

Artificial or synthetic FYM

Synthetic FYM is bulky organic manure resembling ordinary FYM produced without the agency of farm animals.

The process of preparing synthetic FYM from straw was first discovered and patented in England and is known as the ADCO (Agricultural Development Company) (9 % N and 6.5 % H_3PO_4) process of manufacturing artificial FYM. The method is now applicable to all kinds of plant residues and farm waste.

When a plant residue is treated with a solution of nitrogenous compound like Ammonium Sulphate, it undergoes fermentation and gives a product very similar to natural FYM. The addition of phosphate and potash along with soluble nitrogen make the finished product more balance in manural value. Under suitable condition of moisture, air supply and reaction the micro organisms already present in the plant residues bring about a rapid decomposition in the presence o the soluble nitrogen. As a result of which sugars, starch, hemi-celluloses and cellulose of plant resides disappear while lignin, waxes, proteins etc. are converted into humus. The soluble N is fixed or immobilized into insoluble compounds as the protoplasm of the microbial cells. The insoluble N is later ammonified and then nitrified when the manure is applied to the soil.

Each plant residue requires a minimum amount of soluble N to bring about its rapid decomposition. In the case of most straws, which contain 0.4 % N and 45 % carbon, the additional N required is 0.73 parts / 100 parts of the dry material.

$$\left. \begin{array}{l} C :: CN \\ 40 :: 45 \end{array} \right\} \quad \frac{45}{1} \quad \frac{x}{40} \quad 1 = 1.13 \quad 1.13 - 0.40 = 0.73 \text{ N.F.}$$

The additional N so required is known as nitrogen factor. Each plant residue has a different N factor depending on its nitrogen and carbon contents. Ordinarily, a material low in nitrogen has a high nitrogen factor.

In hot countries like ours, it is advisable to substitute Ammonium Sulphate or similar other readily available compounds by a slowly available nitrogenous material like bone-meal, oilcakes or even green leguminous plants. Use of soluble nitrogenous substance leads to a heavy loss of nitrogen during the early stages of decomposition which is considerable reduced when slowly available sources are used.

B. Compost

Compost is composed of organic materials derived from plant and animal matter that has been decomposed largely through aerobic decomposition. The process of composting is simple and practiced by individuals in their homes, farmers on their land, and industrially by industries and cities. Composting is largely a bio-chemical process in which microorganisms both aerobic and anaerobic decompose organic residue and lower the C:N ratio. The final product of composting is well rotted manure known as compost.

Rural compost: Compost from farm litters, weeds, straw, leaves, husk, crop stubble, bhusa or straw, litter from cattle shed, waste fodder, etc. is called rural compost.

Urban compost: Compost from town refuse, night soil and street dustbin refuse, etc is called urban compost.

Composition of urban compost:

Nitrogen	Phosphorus	Potassium
(%N)	(%P_2O_5)	(%K_2O)
1.4	1.0	1.4

Compared to FYM, urban compost prepared from waste and night soil is richer in fertilizer value.

Mechanical Composting Plants

Mechanical composting plants with capacities of 500 – 1000 tonnes per day of city garbage could be installed in big cities in India and 250 tonnes per day plants in the small towns. Refined mechanical compost contains generally about 40% mineral matter and 40% organic materials with organic carbon around 15%. The composition would vary depending on the feed but typically the nutrient content is about 0.7% N, 0.5% P_2O_5 and 0.4% K_2O. There are trace elements like Mn, B, Zn and Cu and the material has C: N ratio of nearly 15-17.

Decomposition

The animal excreta and litter are not suitable for direct use as manure, as most of its manurial ingredients are present in an unavailable form. However urine, if collected separately, can be used directly. The dung and litter have to be fermented or decomposed before they become fit for use. Hence, the material is usually stored in heaps or pits, where it is allowed to decompose. Under suitable conditions of water supply, air, temperature, food supply and reaction, the microorganisms decompose the material. The decomposition is partly aerobic and partly anaerobic. During decomposition the usual yellow or green colour of the litter is changed to brown and ultimately to dark brown or black colour; its structural form is converted into a colloidal, slimy more or less homogenous material, commonly known as humus. A well decomposed manure has a typical black colour and a loose friable condition. It does not show the presence of the original litter or dung.

Factors controlling process of decomposition:

1) **Food supply to micro-organisms and C: N ratio:** The suitable ratio of carbonaceous to nitrogenous materials is 40, if it is wider than this, the decomposition takes place very slowly and when narrow it is quick. C: N ratio of the dung of farm animals varies from 20 to 25, urine 1 to 2, poultry manure 5-10, litters-cereals straw 50, and legume refuse 20.

2) **Moisture:** About 60-70 per cent moisture is considered to be the optimum requirement to start decomposition and with the advance in decomposition, it diminishes gradually being 30-40 per cent in the final product.

   Excess of moisture prevents the temperature form rising high and retards decomposition, resulting in loss of a part of the soluble plant nutrients through leaching and drainage. Hence, in regions receiving heavy rainfall, it is advisable to store the manure or prepare compost in heaps above ground level.

   In the absence of sufficient moisture, microbial activity ceases and the decomposition practically comes to an end.

3) **Aeration:** Most of the microbial processes are oxidative and hence a free supply of oxygen is necessary.

   Reasons for poor aeration in pit/heap

   i) Excessive watering
   ii) Compaction
   iii) Use of large quantities of fine and green material as litters
   iv) High and big heaps or deep pits.

4) Temperature: Under the optimum conditions of air moisture and food supply, there is a rapid increase in the temperature in the manure heap or pit. The temperature usually rises to 50°–60°C and even to 70°C. The high temperature destroys weed seeds, worms, pathogenic bacteria, etc; which prevents fly breeding and makes the manure safe from hygienic point of view.

5) Reaction: The microorganisms liberate certain organic acids during the course of decomposition, which, if allowed to accumulate, retards fermentation and some time even stop it completely. Hence, it is necessary to control the reaction of the material.

A neutral or slightly alkaline reaction between pH 7.0 and 7.5 is considered the most suitable. The addition of alkaline substances like lime and wood ashes neutralized the excess acidity. Since in the preparation of FYM it is a common practice to add household ashes to the manure pit, it is not necessary to add additional alkaline substances.

METHOD OF COMPOSTING

Although utilizing crop wastes in crop production is know from the earliest times systematic work on composting was initiated only in the beginning of this century. In India Howard and Wad (1931) at Indore and Howler (1933) at Bangalore have done some pioneering work. Composting is done either in aerobic condition or in anaerobic condition. Some methods involve both conditions. The advantage of aerobic system is that\t is fast but it require moistening and frequent turning.

Potential of organic and biological resources in India

Sr. No.	Name of Resource	Annual production of biomass (mt)	Nutrient supply (mt)
1.	Cattle and buffaloes (wet dung & urine)	2028.0	6.96
2.	Crop residues	336.0	8.74
3.	Forest litter	100.0	-
4.	City refuse	14.0	0.294
5.	Sewage sludge	6.0	0.011
6.	Press mud	5.0	0.266

Composting is an ancient method by which farmers have been converting plant, animal and human wastes into organic manures in other words we can say value addition to organic wastes. Basically any system or design that ensures efficient decomposition of organic matter can constitute composting method.

1. **Indore method:** Around 1930, Sir Albert Howard, a British Agronomist in India studied composting in a scientific manner at Indore and developed a scientific method of composting, known as Indore method. The waste materials are mixed well and properly moistened with dung or night soil slurry and build into heaps of 4 to 6 m length, 1 m width and 1 m height or put into a pit of 30' x 5'x 3' with slopping sides. In the later method charging of a 30 ft pit done in sections of 5 ft with first section being vacant to facilitate mixing. Aerobic conditions in this method are maintained by periodic manual turning of the composting materials in heaps and mix materials. Water is added if needed. Under this aerobic process, losses of organic matter and nitrogen are to the extent of 40-50% of initial levels. The average composition of manure has been found to be 0.8%N, 0.3% P_2O_5, and 1.5% K_2O.

   Fowler developed the process of **'activated compost'** in which fresh materials were incorporated in an already fermenting heap so that the already established large microbial population could bring about quicker decomposition. This process is useful particularly where offensive materials like night soil are to be quickly and effectively disposed off.

2. **Bangalore method:** Dr. C.N. Acharya in 1938 developed a method for anaerobic composting of city garbage and night soil in pits. The trenches of following dimensions are dug in rows, roads of suitable width are provided between row for the carts to approach and unload the materials inside the trenches.

Population ('000)	Length(m)	Bradth(m)	Depth(m)
<10	4.5	1.5	1.0
10 to 20	6.0	2.0	1.0
20 to 50	9.0	2.0	1.0
>50	10.0	2.5	1.0

The refuse and night soil are spread in alternate layers of 15 cm and 5 cm until the pit is filled 15 cm above ground level, with fine layer of refuse on the top. This may be given a dome shape and covered with a thin layer of soil. The decomposition is mostly anaerobic except in surface layer and is comparatively slow. The C: N ratio is reduced to less than 20:1 in about six months. This method is also known as hot fermentation method as heat loss during decomposition is considerably reduced. Though initially worked out for towns it can be used for making compost from conveniently available organic materials. Under rural conditions, animals dung can be used to substitute night soil.

3. **Coimbatore method:** The composting of wastes is done in pits. A depth of 1.0 m and width of 1.25 m facilitate easy manipulation while filling and turning the material. The length of the pit varies with the quality of material available for composting. A layer of waste material is laid in the pit. It is moistened with sprinkling the slurry of 5-10 kg of cow dung in 2-2.5 lit. of water in which 0.5 to 1.0 kg of fine bone meal is added. Similar layers are laid one over the other till the material rises 0.75 m above ground level. It is plastered over with wet mud and left undisturbed for 8-10 weeks.

   The mud plaster is removed two months later when material is moistened, turned and formed into a rectangular heap in a shady place. It is left undisturbed till required. In the beginning there is an aerobic fermentation when the material is kept covered with mud plaster. Aeration and aerobic fermentation sets in when formed in to an open heap later. The tricalcic phosphate in the bone meal is rendered soluble by the acids produced during decomposition and the compost is enriched by addition of phosphorus. Compost is ready when the temperature in the pile approaches that of the surrounding air. The final product is dark in colour, fairly divided, rich in humus and has a C: N ratio of 10:1 to 20:1.

4. **NADEP method:** This method of composting was developed by a farmer Narayan Dev Rao Panthary Pande of Pusad village in Maharashtra for composting of farm wastes. A structure of bricks with 22 cm thick walls and having size 3.3 m L x 2.0 m W x 1.0 m H is made at high and plain site with the help of cement. For proper aeration 10 cm x10 cm holes are made in wall leaving first and last rows. This structure is called "TANKA" which can use for a longer time. It is not possible to fill up the tanka in one day; therefore, the following materials should be collected in 2-3 days.

   Cow dung 60-100 kg; crop residues, green leaves, stalks, kitchen and fodder wastes etc. 1400-1500 kg; dry sieved soil 1700-1800 kg and water 1500-2000 lit.

   Before filling the tanka, their inner sides are sprinkled with dilute cow dung slurry and then 15-20 cm thick layer of garbage (approximate 200 kg) is laid at the bottom. Second layer is made with cow dung or bio gas slurry by dissolving 4-5 kg cow dung in 150 liter of water. Over this, third layer of dry sieved soil (200-250 kg) free of stone, glasses and plastic etc. is made and water is sprinkled to moisten it. This process is repeated 7-8 times until the material reaches 35-50 cm above the walls. Then it is given a hut shape to check entrance of rain water and plastered with slurry of soil and cow dung. After 45 days the material will compact 25-30 cm downwards and cracks are seen on the surface. Now, the tanka is again filled and plastered in the above manner. In this way 3.5 t of good quality compost can be prepared in 90-110 days.

This method is especially suitable when availability of cow dung is low. Basically, it is an aerobic method in which composting is done in special perforated brick structure to improve the aeration and to minimize the nutrient losses.

Heap V/S Pit decomposition

Heap	Pit
1. Aerobic	1. Anaerobic
2. Turning is required	2. No turning is required
3. Physical disintegration	3. Very little physical disintegration
4. Quick oxidation	4. Slow rate of decomposition
5. High temp. 60° – 70°C. Kill weed seeds and pathogenic organisms	5. High temp. is not developed but weed seeds and MO destroyed due to toxic products of decomposition.
6. Loss of OM is about 50%	6. Loss is about 25%
7. If not properly protected, moisture loss is high. Watering is necessary	7. Moisture loss is minimized. No watering is necessary
8. If rainfall is high, leaching takes place	8. Protected from leaching but anaerobic condition occurs.

C. Vermicompost

Vermicompost is the product of composting utilizing various species of worms, usually red wigglers, white worms, and earthworms to create a heterogeneous mixture of decomposing vegetable or food waste, bedding materials, and vermicast. Vermicast is also known as worm castings, worm humus or worm manure, is the end-product of the breakdown of organic matter by species of earthworm.

The earthworm species (or composting worms) most often used are Red Wigglers (*Eisenia foetida* or *Eisenia andrei*), though European nightcrawlers (*Eisenia hortensis*) could also be used. Users refer to European nightcrawlers by a variety of other names, including *dendrobaenas*, *dendras*, and Belgian nightcrawlers. Containing water-soluble nutrients, vermicompost is a nutrient-rich organic fertilizer and soil conditioner. It provides the vital macro elements such as Nitrogen (0.74%) P_2O_5 (0.97%), K_2O (0.45%) and Ca, Mg and micro elements viz., Fe, Mo, Zn, Cu etc.

Vermiculture means artificial rearing or cultivation of worms (Earthworms) and the technology is the scientific process of using them for the betterment of human beings. Vermicompost is the excreta of earthworm, which is rich in humus.

Earthworms eat cow dung or farm yard manure along with other farm wastes and pass it through their body and in the process convert it into vermicompost. The municipal wastes; non-toxic solid and liquid waste of the industries and household garbage's can also be converted into vermicompost in the same manner. Earthworms not only convert garbage into valuable manure but keep the environment healthy. Conversion of garbage by earthworms into compost and the multiplication of earthworms are simple process and can be easily handled by the farmers.

Methods of vermicompost

In general, following are the three methods of vermicomposting under field condition.

1. Vermicompost of wastes in field pits
2. Vermicompost of wastes in ground heap
3. Vermicompost of wastes in large structures

1. **Vermicomposting of organic wastes in field**

   Pits: It is preferable to go for optimum sized ground pits and 10 x 1.0 x 0.5m (L x W x D) can be effective size of each vermicomposting bed. Series of such beds are to be prepared at one place.

2. **Ground heaps:** Instead of opening of pits, vermicomposting can be taken up in ground heaps. Dome shaped beds (with organic wastes) are prepared and vermicomposting is taken up. Optimum size of ground heaps may be series of heaps of dimension 5.0 x 1.0 x 1.0 m (L x W x H).

3. **Composting in large structures:** vermicomposting is taken up in large structures such as series of rectangular brick columns, cement tanks, stone block etc. which are filled with organic wastes and composting is taken up.

Each of these methods has got advantage as well as limitations. For example in (1) and (3) these would not be any mixing of soil with vermicompost unlike pit system, less incidence of natural enemies. But they need frequent watering (more of labour) compared to pit system. Similarly in places water is scarce (less rainfall tracts); pit system is good which in high rainfall areas (2) and (3) are advantageous as there would be proper drainage.

Steps: This is irrespective of methods

Selection of site: it should be preferably black soil or other areas with less of termite and red ant activity, pH should be between 6 to 8.

Collection of wastes and sorting: for field composting, raw materials are needs in large quantities. The waste available should be sorted in to degradable and non-degradable (be rejected).

Pre-treatment of waste

Lignin rich residues – chopping and subjecting to lignin degrading fungi and later to vermibeds.

Crop stalks and stubbles – dumping it in layers sandwiched with garden soil followed with watering for 10 days to make the material soft and acceptable to worm.

Agro-industrial wastes – mixing with animal dung in 3:1 proportion and later subjecting it for vermicomposting

Insecticidal treatment to site: treating the area as well as beds (in case of pit system) with chlorpyriphos 20 EC @ 3.0 ml/liters to reduce the problem of ants, termites and ground beetles.

Filling of beds with organic wastes: wastes are to filled in the manner given below and each layer should be made wet while filling and continuously watered for next 10 days. In heaping and composting in special structures, the waste is to be dumped serially as done in pits.

7th Layer	A thick layer of mulch with cereal straw	(Top of bed)
6th Layer	A layer of fine soil (Black/garden soil)	(Top of bed)
5th Layer	Dung/FYM/Biogas sludge	(Top of bed)
4th Layer	Green succulent leafy material	(Top of bed)
3rd Layer	Dry crop residues	(Top of bed)
2nd Layer	Dung/FYM/Biogas spent sludge	(Top of bed)
1st Layer	Coconut coir waste/ sugarcane trash	(Bottom of bed)

Except 3rd and 4th layer (which is the material to be degraded) each layer should be 3 to 4 inch thick so that the bed material is raised above the ground level. Sufficient quantity of dry and green wastes is to be used in the beds.

Introduction of worms in to beds: the optimum number of worms to be introduced is 100 No./length of the bed. The species of earthworms that are being used currently for compost production worldwide are *Eisena foetida, Eudirlus eugeniae, Periony excavates, Lumbricus rubella* etc.

Provision of optimum bed moisture and temperature

Bed moisture: by watering at regular intervals to maintain moisture of 60 to 80% till harvest of compost. Temperature requirement for optimal results is 20-30°C by thatching (during summer)

Monitoring for activity of natural enemies and earthworm and management of enemies with botanicals. Promising products: leaf dust of neem, *Acorus calamus* rhizome dust, neem cake etc.

Harvesting of vermicompost and storage

Around 90 days after release of worms, the beds would be ready for harvest.

Stop watering 7days prior to harvest so that worms settle at bottom layer. Collect the compost, shade dry for 12 hours and bag it in fertilizer bags for storage.

Harvest of worm biomass: the worms are to be collected and used for subsequent vermicomposting.

Vermicomposting technique

Sheds: For a vermicomposting unit, whether small or big could be of thatched roof supported by bamboo rafter and purling, wooden trees and stone pillars.

Vermi beds: prepare 90 cm width, 45 cm height and length as per availability of dung and organic waste.

Land: About 0.5 to 1 acre of land will be needed to set up a vermi compost unit cum extension centre.

Seed stock: worms @350 per M^3 of bed space should be adequate to start with and build up the required population in about 2 to 3 cycles.

Water supply system: To maintain optimum moisture content (40%) in vermibed ,spray water on vermin bed. Frequency & quality is regulated by prevailing climatic conditions.

Collection of VC: When vermin compost is ready for collection, top layers apex somewhat dark granular and it used dry tea leaves have been spread over the layer. Watering should then stopped for 2-3 days and ready compost should scrapped form top layers or to a depth.

Storage:-It should be stocked separately in bags. Before packing it should be sieved out from 2 cm galvanized mesh. The compost should not be exposed to sun.

Average nutrient content of vermicompost

1. Organic matter - 30 to 40%
2. Nitrogen - 1.50 to 2.0 %
3. phosphorus - 2.0 to 2.50%
4. potash - 0.6 to 0.80%
5. Ca - 150 to 160 ppm
6. Fe - 120000 to 125000 ppm
7. Zn - 100 to 150 ppm
8. Mn - 200 to 250 ppm
9. Cu - 20 to 30 ppm

Advantages of Vermicomposting

- Vermicompost is an ecofriendly natural fertilizer prepared from biodegradable organic wastes and is free from chemical inputs.
- It does not have any adverse effect on soil, plant and environment.
- It improves soil aeration, texture and tilth thereby reducing soil compaction.
- It improves water retention capacity of soil because of its high organic matter content.
- It promotes better root growth and nutrient absorption.
- It improves nutrient status of soil-both macro-nutrients and micro-nutrients.

Precautions during vermicomposting

- Vermicompost pit should be protected from direct sun light.
- To maintain moisture level, spray water on the pit as an when required.
- Protect the worms from ant, rat and bird

D. NIGHT SOIL

Night soil is manure prepared from human excrement *i.e.* solid and liquid.

Night soil is richer in N, P_2O_5 and K_2O as compared to FYM or compost. On oven dry basis, it has an average chemical composition of:

N%	P_2O_5%	K_2O%
5.5	4.0	2.0

In India it is applied to a limited extent directly to the soil. Pits or trenches of 10 to 12 ft. long, 2 to 3 ft. wide and 9 inches to 1 foot deep are made. In these pits, night soil is deposited and covered over on top with layers of earth or Katchara. This is known as the Poudrette **System**. Since the material formed in the above trenches after they become dry, is known as **poudrette.**

Improved methods of handling night soil:

Since night soil is important bulky organic manure, supplying a good deal of organic matter and plant nutrients to the soil, it is important that night soil is used by the following improved methods:

1. Night soil should be protected from flies and fly breeding should be controlled.
2. It should be stored in such a way that it does not pollute the supply of drinking water.
3. Pathogens, protozoa, cysts, worms and eggs should be destroyed before the night soil is applied to the land.
4. Attempts should be made to compost the night soil with other refuse in urban centers by municipal or town authorities and in rural areas by the farmer himself.

E. Sewage and Sludge

In the modern system of sanitation adopted in cities, water is used for the removal of human excreta and other wastes. This is called the sewage system of sanitation. In this system, there is a considerable dilution of the material in solution and in dispersion in fact, water is the main constituent of sewage, amounting often to 99.0%.

In general sewage has two components, namely

(i) Solid portion, technically known as sludge and

(ii) Liquid portion, commonly known as sewage water.

Both the components are used in increasing crop production as they contain plant nutrients. Both components of sewage as separated and are given a preliminary fermentation and oxidation treatments to reduce the bacterial contamination, the offensive smell and also to narrow down the C:N ratio of the solid portion.

(i) **Sludges:** In the modern system of sewage utilization, solid portion or sludge is separated out to a considerable extent and given a preliminary treatment (i.e. fermentation and oxidation) before its use as manure. Such oxidized sludge is also called **activated sludge** which is of inoffensive smell and on dry weight basis contains up to 3 to 6 per cent N, about 2 per cent P_2O_5 and 1 per cent K_2O in a form that can become readily available when applied to soil.

(ii) **Sewage irrigation:** When raw sewage is treated to remove the solid portion or sludge the water, technically known as **treated effluent**, is used for irrigation purpose. Such a system of irrigation is known as sewage irrigation.

Thus, both the activated sludge and the effluent can be used with safely for manuring and irrigating all field crops except the vegetables which are eaten raw or uncooked.

F. Green Manuring

Practice of incorporating undecomposed green plant tissues into the soil for the purpose of improving physical structure as well as fertility of the soil.

In agriculture, a green manure is a type of cover crop grown primarily to add nutrients and organic matter to the soil. Typically, a green manure crop is grown for a specific period, and then plowed under and incorporated into the soil. Green manures usually perform multiple functions that include soil improvement and soil protection:

- Leguminous green manures such as clover and vetch contain nitrogen-fixing symbiotic bacteria in root nodules that fix atmospheric nitrogen in a form that plants can use.
- Green manures increase the percentage of organic matter (biomass) in the soil, thereby improving water retention, aeration, and other soil characteristics.
- The root systems of some varieties of green manure grow deep in the soil and bring up nutrient resources unavailable to shallower-rooted crops.
- Common cover crop functions of weed suppression and prevention of soil erosion and compaction are often also taken into account when selecting and using green manures.
- Some green manure crops, when allowed to flower, provide forage for pollinating insects.

Historically, the practice of green manuring can be traced back to the fallow cycle of crop rotation, which was used to allow soils to recover.

Types of green manuring:

Broadly two types of green manuring can be differentiated.

i) Green manuring *in situ* and
ii) Green leaf manuring

i) **Green manuring in situ:** In this system green manure crops are grown and buried in the same field, either as a pure crop or as intercrop with the main crop. The most common green manure crops grown under this system are Sannhemp, dhaincha and guar.

ii) **Green leaf manuring:** Green leaf manuring refers to turning into the soil green leaves and tender green twigs collected from shrubs and trees grown on bunds, waste lands and nearby forest areas. The common shrubs and trees used are Glyricidia, Sesbania (wild dhaincha), Karanj, etc.

The former system is followed in northern India, while the latter is common in eastern and central India.

Advantages of Green Manuring:

1. It adds organic matter to the soil. This stimulates the activity of soil micro-organisms.
2. The green manure crops return to the upper top soil, plant nutrients taken up by the crop from deeper layers.
3. It improves the structure of the soil.
4. It facilitates the penetration of rain water thus decreasing run off and erosion.
5. The green manure crops hold plant nutrients that would otherwise be lost by leaching.
6. When leguminous plants, like sunnhemp and dhaincha are used as green manure crops, they add nitrogen to the soil for the succeeding crop.
7. It increases the availability of certain plant nutrients like phosphorus, calcium, potassium, magnesium and iron.

Disadvantages of green manuring:

When the proper technique of green manuring is not followed or when weather conditions become unfavourable, the following disadvantages are likely to become evident.

1. Under rainfed conditions, it is feared that proper decomposition of the green manure crop and satisfactory germination of the succeeding crop may not take place, if sufficient rainfall is not received after burying the green manure crop. This particularly applies to the wheat regions of India.
2. Since green manuring for wheat means loss of kharif crop, the practice of green manuring may not be always economical. This applies to regions where irrigation facilities are available for raising kharif crop along with easy availability of fertilizers.

ORGANIC MANURES

3. In case the main advantage of green manuring is to be derived from addition of nitrogen, the cost of growing green manure crops may be more than the cost of commercial nitrogenous fertilizers.
4. An increase of diseases, insects and nematodes is possible.
5. A risk is involved in obtaining a satisfactory stand and growth of the green manure crops, if sufficient rainfall is not available.

Green manure crops

	Leguminous		Non-leguminous
1.	Sannhemp	1.	Bhang
2.	Dhaincha	2.	Jowar
3.	Mung	3.	Maize
4.	Cowpea	4.	Sunflower
5.	Guar		
6.	Berseem		

Selection of Green manure crops *in situ*

Certain green manure crops are suitable for certain parts of the country. Suitability and regional distribution of important green manure crops are given below:

Sunhemp: This is the most outstanding green manure crop. It is well suited to almost all parts of the country, provided that the area receives sufficient rainfall or has an assured irrigation. It is extensively used with sugarcane, potatoes, garden crops, second crop of paddy in South India and irrigated wheat in Northern India.

Dhaincha: It occupies the second place next to sannhemp for green manuring. It has the advantage of growing under adverse conditions of drought, water-logging, salinity and acidity. It is in wide use in Assam, West Bengal, Bihar and Chennai with sugarcane, Potatoes and paddy.

Guar: It is well suited in areas of low rainfall and poor fertility. It is the most common green manure crop in Rajasthan, North Gujarat and Punjab.

Technique of Green Manuring *in situ*:

The maximum benefit from green manuring cannot be obtained without knowing:
(i) When the green manure crops should be grown,
(ii) When they should be buried in the soil and
(iii) How much times should be given between the burying of a green manure crop and the sowing of the next crop.

(i) **Time of sowing:** The normal practice usually adopted is to begin sowing immediately after the first monsoon rains. Green manure crops usually can be sown/broadcast preferably giving some what higher seed rate.

(ii) **Stage of burying green manure crop:** From the results of various experiments conducted on different green manure crops, it can be generalized that a green manure crop may be turned in soil at the stage of flowering. The majority of the green manure crops take about six to eight weeks from the time of sowing to attain the flowering stage. The basic principle which governs the proper stage of turning in the green manure crops, should aim at maximum succulent green matter at burying.

(iii) Time interval between burying of green manure crop and sowing of next crop.

Following two factors which affect the time interval between burring of green manure crop and sowing of next crop.

1. Weather conditions
2. Nature of the buried green material

In paddy tracts the weather is humid due to the high rainfall and high temperature. These favour rapid decomposition. If the green material to be buried is succulent there is no harm in transplanting paddy immediately after turning in the green manure crop. When the green manure crop is woody, sufficient time should be allowed for its proper decomposition before planting the paddy.

Regions not suitable for green manuring:

The use of green manures in dry farming areas in arid and semiarid regions receiving less than 25 inches of annual rainfall is, as a rule, impracticable. In such areas, only one crop is raised, as soil moisture is limited. Such dry farming areas are located in Punjab, Maharashtra, Rajasthan, M.P. and Gujarat (Kutch and Saurashtra).

On very fertile soils in good physical condition, it is not advisable to use green manures as a part of the regular rotation.

In areas where *rabi* crops are raised on conserved soil moisture, due to lack of irrigation facilities, it is not practicable to adopt green manuring. If green manuring is followed in these areas, there is danger of incomplete decomposition of the green matter and as such less moisture for the succeeding crop.

Plant suitable for green manuring in situ or characteristics of green manures crops:-

An ideal green manures crop should posses the following characteristics as a listed by Agrawal R. R.(1965).

1. It should be a legume with a good nodular growth habit indicative of rapid N fixation under unfavorable soil conditions.
2. It should have little water requirement for its own growth and should be a capable of making good stand on poor and exhausted soils.
3. It should have deeper root system which can open the sub soil and tap lower regions for plant nutrients.
4. The plant should be of a leafy habit capable of producing heavy tender growth early in its life cycle.
5. It should contain large quantities of non-fibrous tissue of rapid decomposability, containing fair percentage.

G. Sheep and Goat Manure

The dropping of sheep and goats contain higher nutrients than farmyardmanure andcompost On an average, the manure contains 3 per cent N, 1 per cent P_2O_5 and 2 per cent K_2O. It is applied to the field in two ways. The sweeping of sheep or goat sheds are placed in pits for decomposition and it is applied later to the field. The nutrients present in the urine are wasted in this method. The second method is sheep penning, wherein sheep and goats are allowed to stay overnight in the field and urine and fecal matter is added to the soil which is incorporated to a shallow depth by running blade harrow or cultivator.

H. Poultry Manure

The excreta of birds ferments very quickly. If left exposed, 50 per cent of its nitrogen is lost within 30 days. Poultry manure contains higher nitrogen and phosphorus compared to other bulky organic manures. The average nutrient content is 3.03 per cent N, 2.63 per cent P_2O_5 and 1.4 per cent K_2O.

7.2.2. Concentrated Organic Manures

Concentrated organic manures are those that are organic in nature, plant or animal origin and contain higher percentage of major plant nutrients like N, P, and K compared to bulky organic manures like FYM and compost. The common concerted organic manures are oilcake, blood-meals, fish manure, meat meal and wool waste.

A. Oil Cake

Oil cake is the residues left after the oil is extracted from the oil bearing seeds.

It contains varying quantities of oil depending upon the process of manufacture employed in treating the oilseed as shown below.

Process of manufacture	Oils %
Country ghani	10 -15
Hydraulic press	8 -10
Expeller	5 -8
Solvent	1 -2

Oil cake can be grouped into tow classes

1. **Edible oilcake**:- Suitable for feeding to cattle as concentrates and are seldom used as manure, except when it becomes mouldy or rancid and found unfit for feeding cattle. eg. mustard oil cake, groundnut oil cake, sesame or til cake, linseed cake, coconut cake etc.
2. **Non-edible oilcake**:-This type of oilcake is not suitable for feeding to cattle and mainly used for manuring crops. eg. castor cake, neem cake, mahuva cake etc. The non-edible oilcake contains a harmful or toxic substance which makes them unsuitable for feeding to cattle.

Characteristics of oil cakes

1. Oilcakes are quick acting organic manure as C:N ratio is usually narrow (5-15:1). Though they are insoluble in water, their N becomes quickly available to the plants in about a week or 10 days after its application.
2. Oil prevents rapid conversion of nitrogen. Country ghani or gana oilcakes usually contain a little more oil than or expeller-pressed oilcakes. Owing to the higher percentage of oil, country ghani oilcakes are somewhat slow acting.
3. Nearly 50-80 % of its N is made available in 2-3 months and rate of availability varies with type of cake and nature of soil.
4. Castor cake contains "Ricin", mahuva cake contains "saponin" and neem cake contains "nimbidin" which are responsible for slow nitrification of their N due to effect on soil micro-organisms.
5. Castor cake has also good vermicidal effect against white ants.
6. Groundnut cake has the highest nitrification rate.
7. Mahuva cake is very poor in N content and take a long time to nitrify. As such, mahuva cake should be applied about 2 to 3 months before sowing/planting of the crop.

Precautions in using the oilcakes

1. Oilcakes should be well powdered before application, so that they can be spread evenly and are easily decomposed by micro-organisms.
2. They can be applied a few days prior to sowing or at the time of sowing.
3. It is best suited as a top dressing after the plants have established themselves.
4. Use only when there is sufficient moisture in the soil or restrict use to irrigated crops or in tracts having sufficient high rainfall.
5. If mahuva cake is to be used, apply before 2-3 months before planting or decompose in a pit and then apply or treat with ammonium sulphate.

Average nutrient contents of principle oilcakes.

Name of the oilcake	Percentage composition		
	N	P_2O_5	K_2O
Non-edible oilcake			
Castor cake	4.3	1.8	1.3
Cotton seed cake (Undecorticated)	3.9	1.8	1.6
Karanj cake	3.9	0.9	1.2
Mahuva or ippi cake	2.5	0.8	1.8
Neem cake	5.2	1.0	1.4
Safflower cake(Undecorticated)	4.9	1.4	1.2
Edible oil cake			
Coconut cake	3.0	1.9	1.8
Cotton seed cake (Undecorticated)	6.4	2.9	2.2
Groundnut cake	7.3	1.5	1.3
Linseed cake	4.9	1.4	1.3
Rapeseed cake	5.2	1.8	1.2
Safflower cake(Undecorticated)	7.9	2.2	1.9
Sesame or til cake	6.2	2.0	1.2

B. Blood Meal

Dried blood or blood meal contains 10 to 12 % N, 1.0 to 2.0 % P_2O_5 and 1.0 % K_2O. Blood meal is prepared from the blood collected from slaughter house treated with $CuSO_4$, dried, powdered, bagged and sold as blood meal. Blood meal is quick acting organic manure and is effective for all crops on all soils. It should be applied like oilcakes.

C. Meat Meal

Bones and meat are cooked in special types of pan for 2-3 hours. Bones are separated and meat is dried and powdered. It is Quick acting and used like oilcake. It contains 10.5 % N and 2.5 % P_2O_5.

D. Fish Manures

Non-edible fish and fish waste is dried and powdered. It is quick acting organic manure and used like oilcakes for all crops on all types of soils. Fish meal or fish manure contain 4-10 % N, 3-9 % P_2O_5 and 0.3-1.5 % K_2O

E. Horn and Hoof Meal

Horn and hoof cooked in bone digester, dried and powdered which contains about 13 % N.

F. Night Soil

Night soil is the human excrement-solid and liquid. It is rich in nutrient contain. On an average, it contains 1.0-1.6 % N, 0.8 % P and 0.2-0.6 % K on over dry basis. It is a good manure for production of crop. But there is some prejudice of using night soil for crop production. In India, it is directly applied to the soil to a limited extent.

G. Guano

The material obtained from the excreta and bodies of sea birds are known as guano. It contains 7-8 % N, 11-14 % P and 2.3-3.0 % K.

7.3. ROLE OF ORGANIC MANURES

1. When the coarse organic matter applied on soil surface, reduce the impact of the falling rain drop and permits clear water to seep gently into the soil. Surface run-off and erosion are thus reduced and as a result, there is more available water for plant growth.
2. The addition of easily decomposable organic residues causes the synthesis of complex organic substances that bind soil particles into structural unit called aggregates. These aggregate helps to maintain lose, open granular conditions. Water is then able to enter and percolate more rapidly downward through the soil. The granular condition of soil maintains favorable condition of aeration and permeability.
3. Water holding capacity is increased by organic matter. The fact that it does not necessarily mean an increase in available water supplies to plant, since organic matter holds water fairly, tightly, thus the PWP is increased. Organic

matter definitely increases the amount of available water in sandy and loamy soils. Further the granular soil resulting from organic matter additions, supplies more water than sticky and impervious soil.

4. Live roots decay and provide channels down in soil through which new plant roots grow more luxuriantly. The same roots channels are effective in transmitting water downward, a part of which is stored for further use by plants.

5. Evaporations losses of water are reduced by organic mulches.

6. Trashy, coarse organic matter on the surface of the soil reduces the losses of soil by wind erosion.

7. Surface mulches lower the soil temperature in summer and keep the soil warmer in winter.

8. Organic matter servers as a reservoir of chemical elements that are essential for plant growth. Most of the soil N occurs in organic combination. Only, a small traction, usually 1-3 %, occurs in inorganic forms at nay one time. Also a considerable quantity of P and S exits in organic forms. Upon decomposition, organic matter supplied the nutrients needed by growing plants as well as many hormones and antibiotics. These are released in harmony with the needs of the plants. When the environmental conditions are favorable for rapid plant growth, the same condition favour in the rapid release of nutrients from the organic matter.

9. Organic matter upon decomposition produces organic acids and CO_2 which helps to dissolve minerals such as K and to make more available to the growing plants.

10. Organic matter helps to buffer soils against rapid chemical changes in pH due to addition of lime and fertilizers.

11. Humus provides a storehouse for the exchangeable and available cations – K^+, Ca^{++} and Mg^{++}. Ammonium fertilizers are also prevented from leaching because humus holds ammonium in an exchangeable and available form.

12. Fresh organics matter has a special function in making soil P more rapidly available in acid soil. Upon decomposition, organic matter release citrates, oxalates, tart rates and lactates which combined with iron and aluminum more rapidly than does phosphorus. This results in the formation of less amounts of insoluble iron and aluminums phosphates and availability of more P for plant growth.

13. Organic acids released from decomposition of organic matter helps to reduce alkalinity in soils.

14. Organic manure increases the cation exchange capacity (CEC). Thus, it prevents the loss of nutrient by leaching and retains them in available form.

122 SOIL FERTILITY AND NUTRIENT MANAGEMENT

15. The organic matter serves as a source of energy for the growth of soil microorganisms. All heterotrophic organisms eg. Nitrogen fixing organisms requires easily decomposable organic matter as their source of carbon. Without carbon, nitrogen fixation by Azotobactor and Clostridium would be impossible.

16. Fresh organic matter supplies food for soil life such as earthworms, ants, and rodents. These animal burrow channels in soil and construct extensive channels through the soil that serve not only loosen the soil but also to improve drainage and aeration. Further, this permits plant roots to obtain oxygen and to release CO_2 as they grow. Earthworms can flourish only in soils that are well provided with organic matter.

7.4. TRANSFORMATION OF ORGANIC MANURES

Organic matter in the soil comes from the remains of plants and animals. As new organic matter is formed in the soil, a part of the old becomes mineralized. The original source of the soil organic matter is plant tissue. Under natural conditions, the tops and roots of trees, grasses and other plants annually supply large quantities of organic residues. Thus, higher plant tissue is the primary source of organic matter. Animals are usually considered secondary sources of organic matter. Various organic manures, that are added to the soil time to time, further add to the store of soil organic matter.

Composition of plant residues

Composition of organic residues have un-decomposed soil organic matter (mainly plant residues together with animal remains, i.e. animal excreta etc.) The moisture

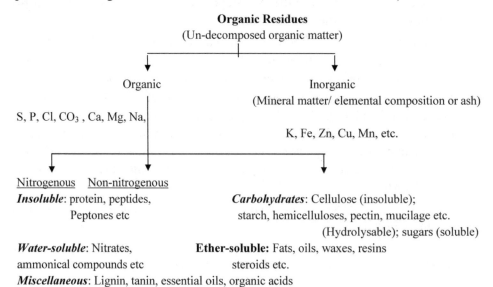

content of plant residues varies from 60 to 90% (average 78%) and 25% dry matter (solid). Plant tissues (organic residues) may be divided into 91) organic and (2) inorganic (elemental) composition. The compounds constituting the plant residues or un-decomposed soil organic matter is shown in the following diagram

Transformation reaction of organic manures in soil

The organic materials incorporated in the soil do not remain as such very long. They are at once attacked by a great variety of microorganisms, worms and insects present in the soil especially if the soil is moist. The microorganism for obtaining their food, break up the various constituents of which the organic residues are composed, and convert them into new substances, some of which are very simple in composition and others highly complex. The whole of the organic residues is not decomposed all at once or as a whole. Some of the constituents are decomposed very rapidly, some less readily, and others very slowly.

A tentative scheme for the different stages of microbial decomposition of organic residues are shown in Fig. 7.1.

It is evident that different constituents of organic residues decompose at different rates. Simple sugars, amino acids, most proteins and certain polysaccharides decompose very quickly and can be completely utilized within a very short period. Large macro-molecules which make up the bulk of plant residues must first be broken down into simpler forms before they can be utilized further for energy and cell synthesis. This process is carried out by certain specific enzymes excreted by microorganisms.

The utilization of residue componenets and their brokendown products (sugar, aminoacids, phenolic compounds and others) leads to the production of microbial cells, which are further degrades following death of organisms.

Importance of C:N ratio in rate of decomposition

The ratio of the weight of organic carbon (C) to the weight of total nitrogen (N) in a soil (or organic material), is known as C: N ratio. When fresh plant residues are added to the soil, they are rich in carbon and poor in nitrogen. The content of carbohydrates is high. This results in wide carbon-nitrogen ratio which may be 40 to 1. Upon decomposition the organic matter of soils changes to humus and have an approximate C: N ratio of 10:1.

The ratio of carbon to nitrogen in the arable (cultivated) soils commonly ranges from 8:1 to 15:1. The carbon-nitrogen ratio in plant material is variable, ranging from 20:1 to 30:1. Low ratios of carbon to nitrogen (10:1) in soil organic matter generally indicate an average stage of decomposition and resistance to further microbiological decomposition. A wide ratio of C: N (35:1) indicates little or no decomposition, susceptibility to further and rapid decomposition and slow nitrification.

124 SOIL FERTILITY AND NUTRIENT MANAGEMENT

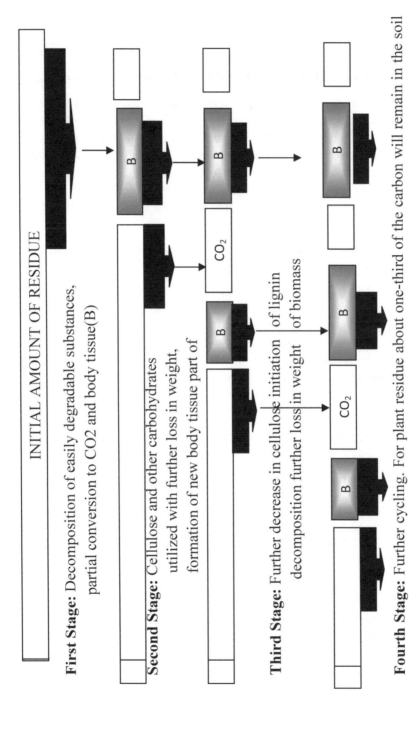

Fig.7.1: Stage of decomposition of organic residues

Significance of C:N Ratio

(1) Keen competition for available nitrogen when organic residues (with high C:N ratio) are added to soils. When organic residues with a wide C/N ratio (50:1) are incorporated in the soil, decomposition quickly occurs. Carbon dioxide is produced in large quantities. Under these conditions, nitrate-nitrogen disappears from the soil because of the instant microbial demand for this element to build up their tissues. And for the time being, little (or no) nitrogen is an available to plants. As decomposition occurs, the C/N ratio of the plant material decrease since carbon is being lost and nitrogen conserved. Nitrates-N again appear in quantity in the soil, thus, increases plant growth.

(2) *Consistency of C: N Ratio.* As the decomposition processes continue, both carbon and nitrogen are now subject to loss the carbon as carbon dioxide and the nitrogen as nitrates which are leached or absorbed by plants. At a point carbon-nitrogen ratio, becomes more or less constant, generally stabilizes at 10:1 or 12:1.

7.5. BIO FERTILIZERS

Biofertilizers are defined as preparations containing living cells or latent cells of efficient strains of microorganisms that help crop plants uptake of nutrients by their interactions in the rhizosphere when applied through seed or soil. They accelerate certain microbial processes in the soil which augment the extent of availability of nutrients in a form easily assimilated by plants.

Very often microorganisms are not as efficient in natural surroundings as one would expect them to be and therefore artificially multiplied cultures of efficient selected microorganisms play a vital role in accelerating the microbial processes in soil.

Use of biofertilizers is one of the important components of integrated nutrient management, as they are cost effective and renewable source of plant nutrients to supplement the chemical fertilizers for sustainable agriculture. Several microorganisms and their association with crop plants are being exploited in the production of biofertilizers. They can be grouped in different ways based on their nature and function.

Sr. No.	Groups	Examples
N_2 Fixing Biofertilizers		
1.	Free- living	Azotobactor, Beijerinkia, Clostridium, Klebsiella, Anabaena, Nostoc
2.	Symbiotic	Rhizobium, Frankia, Anabaena azolla

[Table Contd.

Contd. Table]

Sr. No.	Groups	Examples
3.	Associative Symbiotic	Azospirillum
P Solubilizing Biofertilizers		
1.	Bacteria	*Bacillus megaterium var. phosphaticum, B. subtilis, B. circulans, pseudomonas striata*
2.	Fungi	*Penicillium sp, Aspergillus awamori*
P Mobilizing Biofertilizers		
1.	Arbuscular mycorrhiza	*Glomus sp., Gigaspora sp., Acaulospora sp., Scutellospora sp. & Sclerocystis sp.*
2.	Ectomycorrhiza	*Laccaria sp., Pisolithus sp., Boletus sp., Amanita sp.*
3.	Ericoid mycorrhizae	*Pezizella ericae*
4.	Orchid mycorrhiza	*Rhizoctonia solani*
Biofertilizers for Micro nutrients		
1.	Silicate and Zinc solubilizers	*Bacillus sp.*
Plant growth promoting Rhizobacteria		
1.	Pseudomonas	*Pseudomonas fluorescens*

NITROGEN FIXING BIOFERTILIZERS

(a) Symbiotic nitrogen fixers

1. ***Rhizobium:*** Among all bio-fertilizers, *rhizobium* inoculants are widely used by farmers throughout the world. These organisms colonize roots of leguminous plants to form root nodules, which act as the factories of nitrogen production for the host plant. These bacteria live in these nodules and takes nitrogen of air to convert into an organic form that the plant can use. As the bacteria live right in root it transfers nutrients directly into the plants. *Rhizobium*- legume symbioses can fix 100-300 kg N/ha on a season depending upon crop and live substantial quality of N in surroundings rhizosphere for the succeeding crop.

   In chick pea, nitrogen fixation starts about 15 days after sowing when nodules are small but pink and attain peak level at the time of flowering and early stage of seed formation. Because of this fact it is advisable to give starter dose of 20-25 kg N/ha to legume crops. *Rhizobium* can meet more than 80% of N need of the legume crops with 10-25% increase of grain yield of pulses. The response varies depending on soil conditions and effectiveness of native population.

2. ***Azolla:*** Azolla is a floating water fern and is ubiquitous in distribution. It has an algal symbiont viz., Anabaena Azollae within its central cavity. The alga

fixes atmospheric nitrogen and is present at all stages of growth and development of fern. Azolla contains 0.2-0.3 % N on fresh weight basis and 3-5 % on dry weight basis. It's used as bio-fertilizer for rice in many countries and relatively more advantageous over urea. Under ideal conditions, it has potential of fixing more than 10 kg N/ha/day. One crop of Azolla provides 20-40 kg N/ha to the rise in about 20-25 days. Farmers can take two such crops during rice cultivation. Technology of *A. pinnata* cultivation is developed and well domesticated, which paddy growers could easily adopt. Azolla fern has greater potential as sole crop in specific areas viz; low land paddy, water logged waste lands, seepage water, shallow ponds, natural fresh water lagoons, burrow pits, Khet talawadi etc. and thereby it generates employment in rural areas. Over all, Azolla is widely accepted as fertilizer, feed, food and fodder.

(b) Non Symbiotic Nitrogen Fixers

1. *Azotobacter*: These are free-living gram-negative rod shaped nitrogen-fixing bacteria in loose association for plants. They are normal inhabitants of soils, ubiquitous in geographic distribution an can a variety of C and energy for their growth. These bacteria can substitute 20-40 kg N/ha for different crops. In India, depending upon soil fertility most probable number (MPN) of N fixers slime, which helps in soil aggregation. Seed germination and plant stand are improved in plants upon inoculation with improved strains. Various species of *Azotobacter* are *A. agilis, A. chrococcum, A. beijerinckii, A.vinelandi, A. ingrinis*. Out of these A. chrococcum happens to be the dominant inhabitant in arable soils and is most effective and widely used.They increase crop yield by 10-15% and help in mineralization of plant nutrients and proliferation of other useful micro-organism.

2. *Azospirillum* : *Azospirillum* are non-symbiotic N fixing bacteria. This is very important and widely used biofertilizer in present day agriculture for many crops. These bacteria have intimate association with roots of cereals and grasses. Individual cells are gram-negative curved rods, 1 mm in diameter, size and shape vary. There are four common species viz; *A. lipoferrum, A. brazillense, A. amazonense* and *A. halpraeferans*. The mechanism by which inoculated plants derive positive benefits is same as *Azotobacter* and fix 20-40 kg N/ha in field conditions with increase in yield by 10-15 %.

3. *Acetobacter*: This is a sacharophillic bacteria and associate with sugarcane, sweet potato and sweet sorghum plants and fixes 30 kg N/ha/year. Mainly this bacteria is commercialized for sugarcane crop. It is known to increase yield by 10-20 t/ha and sugar content by about 10-15%.

Phosphate Solubilizing Biofertilisers

PSMs includes different group of microorganisms such as bacteria, yeast and fungi, which convert insoluble inorganic phosphates into soluble form. The common genera like *Bacillus, Pseudomonas, Aspergillus, Penicillium, , Fusarium, Micrococcus* etc have been reported to be active in bioconversion of PO_4. It is estimated that in most tropical soils only 25% is available for plant growth and 75% of super phosphate applied get fixed. Important species of PSM includes *Bacillus polymaxa, B. coagulans, B. circulans, Psuedomonas striata, Aspergillus awamori and Penicillum digitatum.* PSMs can be mass multiplied on Pikovasky's broth and mixed with carrier material for field use. These organisms possess the ability to bring phosphate solubilization by secreting organic acids such as formic, acetic acids, propionic, lactic, glycolic, succinic acids etc. These acids lower the p^H and bring about the dissolution of bound form of phosphates. The integrated use of PSMs could bring benefits from the low-grade rock phosphates available to the tune of 230 million tons in our country. PSMs are recommended for all crops in India and have shown to replace 20-50 kg P_2O_5/ha in different crops due to inoculative applications.

Liquid Biofertilizers – A New Panorama

Successfully developed **Anubhav liquid formulations of bio-fertilizers** based on native Azotobactor and Phosphate culture, product is having minimum cell count of 10^9/ml and with shelf life of 1 year, suitable for drip irrigation and green house cultivation as against currently marketed carrier based (lignite) products having shelf life of 6 months. Following demonstrations at farmers' field in tribal areas of Gujarat in maize, wheat, mung etc. during last decade, **Lab to Land,** Showed saving up to 50% N+P with significant yield increase. Sale of liquid bio-fertilizers to the end users since 2005 is up to 3000 lit at affordable price Rs. 100/-lit. This technology is ready to be transferred to the farming community using concept of public private partnership.

Dosage of liquid Bio-fertilizers in different crops

Recommended liquid Bio-fertilizer and its application method, quantity to be used for different crops are as follows:

ORGANIC MANURES

Crop	Recommended Bio-fertilizer	Application method	Quantity to be used
Field crops, pulses, Chickpea, Pea, Groundnut, Soybean, Beans, Lentil, Lucern, Berseem, Green gram, Black gram, Cowpea and pigeon pea	*Rhizobium*	Seed treatment	500 ml/acre
Cereals, Wheat, Oat, Barley	*Azotobactor / Azospirillum*	Seed treatment	500 ml/acre
Rice	*Azospirillum*	Seed treatment	500 ml/acre
Oil seeds, Mustard, Seasum, Linseeds, Sunflower, Castor	*Azotobacter*	Seed treatment	500 ml/acre
Millets, Pear millets, Finger millets, Kodo millet	*Azotobacter*	Seed treatment	500 ml/acre
Maize and Sorghum	*Azospirillum*	Seed treatment	500 ml/acre
Forage crops and Grasses Bermuda grass, Sudan grass, Napier Grass, Para Grass, Star Grass etc.	*Azotobactor*	Seed treatment	500 ml/acre
Other Misc. Plantation Crops Tobacco	*Azotobactor*	Seedling treatment	1250 ml/acre
Tea, Coffee	*Azotobacter*	Soil treatment	400 ml/acre
Rubber, Coconuts	*Azotobactor*	Soil treatment	2-3 ml/plant
Agro-Forestry/Fruit plants All fruit/agro-forestry (herb, shrubs, annuals and perennial) plants for fuel wood fodder, fruits, gum, spice, leaves, flowers, nuts and seed purpose	*Azotobacter*	Soil treatment	2-3 ml/plant at nursery
Leguminous plants/trees	*Rhizobium*	Soil treatment	1-2 ml/plant

Note: Doses recommended when count of inoculums is 1×10^8 cells/ml then doses will be ten times more besides above said Nitrogen fixers, Posphate solubilizers and potash mobilizers at the rate of 200 ml/acre could be applied for all crops.

Application of Bio-fertilizers

1. Seed treatment or seed inoculation
2. Seedling root dip
3. Main field application

Seed Treatment: One packet of the inoculant is mixed with 500 ml of rice kanji to make a slurry. The seeds required for ha are mixed in the slurry so as to have a uniform coating of the inoculants over the seeds and then shade dried for 30 minutes. The shade dried seeds should be sown within 24 hours. One packet of the inoculants (200 g) is sufficient to treat 10 kg of seeds.

Seedling root dip: This method is used for transplanted crops. Five packets of the inoculants is mixed in 100 liters of water. The root portion of the seedlings required for one ha is dipped in the mixture for 5 to 10 minutes and then transplanted.

Main field application: Ten packets of the inoculants is mixed with 50 kgs of dried and powdered farm yard manure and then broadcasted in one hectare of main field just before transplanting.

Rhizobium: For all legumes *Rhizobium* is applied as seed inoculants.

Azospirillum/Azotobacter: In the transplanted crops, *Azospirillum* is inoculated through seed, seedling root dip and soil application methods. For direct sown crops, Azospirillum is applied through seed treatment and soil application.

Phosphobacteria: Inoculated through seed, seedling root dip and soil application methods as in the case of *Azospirillum*.

Combined application of bacterial biofertilizers: Phospho bacteria can be mixed with *Azospirillum* and *Rhizobium*. The inoculants should be mixed in equal quantities and applied as mentioned above.

Points to Remember

- Bacterial inoculants should not be mixed with insecticide, fungicide, herbicide and fertilizers.
- Seed treatment with bacterial inoculants is to be done at last when seeds are treated with fungicides.

ORGANIC MANURES 131

1 kg. bio-fertilizer in
50 litres of water

Seedlings

Dipping seedlings
for 30 min.

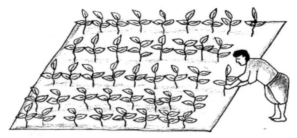
Transplanting the seedlings

Seedling Dip

Bio-fertilizer (5 kgs)

Mixing bio-fertilizer with 100
kgs of farm yard manure
(FYM)

Bio-fertilizer mixture ready for use

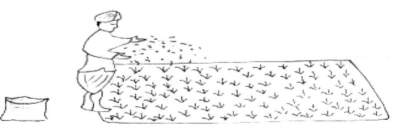
Spreading the mixture in the field

Fig. 7.2: Soil application of bio-fertilizers

SOIL FERTILITY AND NUTRIENT MANAGEMENT

Bio-fertilizers recommendation (one packet- 200 g)

Sr. No.	Crops	Seed	Nursery	Seedling dip	Main field	Total requirement of packets per ha
1.	Rice	5	10	5	10	30
2.	Sorghum	3	–	–	10	13
3.	Pearl millet	3	–	–	10	13
4.	Rangi	3	–	5	10	18
5.	Maize	3	–	–	10	13
6.	Cotton	3	–	–	10	13
7.	Sunflower	3	–	–	10	13
8.	Castor	3	–	–	10	13
9.	Sugarcane	10	–	–	36(split)	46
10.	Turmeric	–	–	–	24(split)	24
11.	Tobacco	1	3	–	10g/pit	14
12.	Papaya	2	–	–	10	–
13.	Mandarin orange	2	–	–	10 g/pit	–
14.	Tomato	1	–	–	10	14
15.	Banana	–	–	5	10 g/pit	–

Rhizobium **(only Seed application is recommended)**

Sr. No.	Crop	Total requirement of packets per ha
1.	Soybean	5
2.	Groundnut	5
3.	Bengal gram	5
4.	Black gram	3
5.	Green gram	3
6.	Red gram	3
7.	Cow pea	3

Phosphobacteria: The recommended dosage of *Azospirillum* is adopted for phosphor-bacteria inoculation; for combined inoculation, both bio-fertilizers as per recommendations are to be mixed uniformly before using.

CHAPTER 8

CHEMICAL FERTILIZERS

"Fertilizer may be defined as materials having definite chemical composition with a higher analytical value and capable of supplying plant nutrients in available forms."

Usually fertilizers are inorganic in nature and most of them are the products of different industries. Only exception to the inorganic nature. Urea and $CaCN_2$ (calcium cynamide) and solid organic nitrogenous fertilizers. Required in less quantity concentrated and cheaper. Nutrients are readily available. Very less residual effect. Salt effect is high. Adverse effects are observed when not applied in time and in proper proportion.

Complete Fertilizer: Complete fertilizer is referred to a fertilizer material which contains all three major nutrients, N, P and K.

Incomplete Fertilizers: This fertilizer is referred to a fertilizer material which lacks any one of three major nutrient elements.

Straight Fertilizer: Straight fertilizers may be defined as chemical fertilizers which contain only one primary or major nutrient element. e.g. ammonium sulphate,$(NH_4)_2SO_4$, Urea$(CO(NH_2)_2)$.

Complex fertilizer: This fertilizer may be defined as a fertilizer material which contains more than one primary or major nutrient elements produced by the process of chemical reactions.

There are generally three types of chemical fertilizers available in the market namely nitrogenous, Phosphatic and Potassic fertilizers. These types of fertilizers are mostly used by the Indian farmers for the crop cultivation. The most important nitrogenous, Phosphatic and Potassic fertilizers used by the farmers are ammonium sulphate, urea as N sources, superphosphate and rock phosphate as P sources and muriate of potash as K sources. Sometimes micronutrient fertilizers like $ZnSO_4$ as Zn, Borax as B-sources etc. are used by the farmers.

8.1 NITROGENOUS FERTILIZERS

Nitrogen is present in soil as (i) Organic form and (ii) inorganic form. Inorganic form includes Ammonical (NH_4^+), Nitrite (NO_2^-) and Nitrate (NO_3^-). Plant absorbs N in the form of NO_3^- and NH_4^+ forms by paddy in early stages. Nitrogen in NH_4^+ form goes on exchange complex on clay and organic colloids and hence, this part is not lost due to leaching, while NO_3^- is lost due to leaching as it does not go on exchange complex under neutral to higher pH values of soil. But it goes on exchange under highly acidic conditions. The nitrate fertilizers are hygroscopic in nature, it is for this reason, nitrate fertilizers are not commonly used even though plant absorbs N as NO_3^-. Therefore, organic form (urea) and fertilizers of NH_4 form like ammonium sulphate are widely used.

Most of Indian soils are low in N and the requirement of N by crop is throughout its growing period, therefore N should be applied in such a way that plant gets it throughout its life period. It becomes absolutely necessary to apply nitrogenous fertilizers to every soil and crop. For this, the total quantity of nitrogenous fertilizers requirement is more compared to fertilizers of other nutrients.

Commercial Nitrogenous Fertilizers: Commercial nitrogenous fertilizers are those fertilizers that are sold for their nitrogen content and are manufactured on a commercial scale.

Classification of Nitrogenous Fertilizers: Nitrogenous fertilizers may be classified into four groups on the basis of the chemical form in which nitrogen is combined with other elements with a fertilizer.

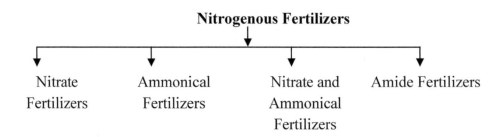

1) **Nitrate Fertilizers**: Nitrogen is combined as NO_3^- with other elements. Such fertilizers are:
   i) Sodium nitrate or Chilean nitrate ($NaNO_3$) – 16% N
   ii) Calcium nitrate [$Ca(NO_3)_2$] – 15.5% N.
   Out of these, sodium nitrate is an imported commercial fertilizer.

2) **Ammonical Fertilizers:** In these fertilizers, nitrogen is combined in Ammonical (NH_4) form with other elements. Such fertilizers are
   i) Ammonium sulphate [$(NH_4)_2 SO_4$] – 20% N
   ii) Ammonium Chloride (NH_4Cl) - 24 to 26% N
   iii) Anhydrous ammonia - 82% N

3) **Nitrate and Ammonical Fertilizers:** These fertilizers contain nitrogen in the form of both nitrate and Ammonical. Such fertilizers are
   i) Ammonium nitrate ($NH_4 NO_3$) - 33 to 34% N
   ii) Calcium ammonium nitrate - 26% N
   iii) Ammonium sulphate nitrate - 26% N

4) **Amide fertilizers:** These fertilizers contain nitrogen in amide or cynamide form. Such fertilizers are:
   i) Urea [$CO(NH_2)_2$] – 46% N
   ii) Calcium cynamide ($CaCN_2$) – 21% N

General Properties of Nitrogenous Fertilizers

1) **Nitrate fertilizers:** Most of the field crops except paddy in early stages of their growth, take up nitrogen in nitrate form as such,
   i) Nitrate fertilizers are readily absorbed and utilized by these crops. Nitrate fertilizers are very often used as top and side dressings.
   ii) The great mobility of the nitrate ion in the soil has the advantage that, even by broadcasting the fertilizer on the surface of the soil, the nitrogen reaches the root zone quickly.
   iii) On the other hand, there is also the increased danger of leaching of these fertilizers. On dry soils, nitrate fertilizers are superior to the other forms of nitrogenous fertilizers.
   iv) All nitrate fertilizers are basic in their residual effect on the soils and their continued use may play a significant role in reducing soil acidity. Sodium nitrate, for example, has a potential basicity of 29 pounds of calcium carbonate per 100 pounds of fertilizer material.

136 SOIL FERTILITY AND NUTRIENT MANAGEMENT

2) **Ammonical Fertilizers:**
   i) Ammonical fertilizers are water soluble.
   ii) It is less rapidly used by plant than NO_3^-, as it is to be changed to NO_3^- before use by crop.
   iii) It is resistant to lost due to leaching as being cation goes on exchange complex.
   iv) Any fertilizers which contain N as NH_4^+ or which is changed as NH_4^+ produced acidity in soil due to production of HNO_3.
   v) Ammonium (NH_4^+) of fertilizer goes on exchange complex, used by crop like paddy.
   vi) Used by microorganisms nitrified to NO_3 and lost due to volatilization from soil.

3) **Nitrate and Ammonical Fertilizers:**
   i) Fertilizers of this group are soluble in water.
   ii) Nitrate part can readily be used by crop.
   iii) NH_4^+ can go on exchange and hence, this is best type but did not over take ammonium sulphate and urea, as they are hygroscopic in nature.
   iv) They are acidic in their residual effect on soil

4) **Amide Fertilizes:**
   i) Fertilizers of this group are readily soluble in water. They are easily decomposed by microorganisms due to presence of oxidisable carbon.
   ii) They are quickly changed to NH_4^+ then in to NO_3^-.

Manufacturing process of ammonium sulphate and urea

Most of the nitrogenous fertilizes like ammonium sulphate, urea, ammonium nitrate, ammonium sulphate nitrate and even DAP are manufactured by using Anhydrous Ammonia gas (NH_3) as one of the important compound. Most of the commercial NH_3 is prepared by Haber's process by the fixation of atmospheric N by means of H_2.

The reaction is:

$$N_2 + 3H_2 \xrightarrow[\text{at 550°C temp} \atop \text{Fe and Mo as catalysts}]{\text{200 atm. Pressure}} 2NH_3 + 24.4 \text{ KCal}$$

CHEMICAL FERTILIZERS 137

Ammonia can also be obtained from natural gas, coal gas and naphtha. Therefore, cost of fertilizer production in fertilizer factory installed near a petrochemical will be low.

The NH_3 gives ammonium sulphate with sulphuric acid, NH_4Cl with HCl; NH_4NO_3 with HNO_3; urea with CO_2; MAP and DAP with H_3PO_4. Thus, NH_3 is chief compound for most of the nitrogenous fertilizers.

i) Preparation of Ammonium sulphate (A/S):-

It is prepared by

(a) Reacting NH_3 with H_2SO_4

(b) Gypsum process

(c) By-product of coal and steel industries.

a) NH_3 with H_2SO_4 :- NH_3 is reacted with H_2SO_4 giving A/S. The liquid is crystallized and crystals of A/S are obtained.

$$2NH_3 + H_2SO_4 = (NH_4)_2SO_4$$

Since the sulphur used in sulphuric acid is to be imported, the source of H_2SO_4 becomes costlier and hence, gypsum a cheaper source of sulphur is used in gypsum process.

b) Gypsum process: The main raw materials required in gypsum process are NH_3, pulverized gypsum, CO_2 and water. NH_3 is obtained by Haber's process. This NH_3 when reacts with CO_2 gives $(NH_4)_2CO_3$. The ground gypsum when reacts with $(NH_4)_2CO_3$ solution gives $(NH_4)_2SO_4$ and $CaCO_3$. The reactions are :

$$N_2 + 3H_2 \longrightarrow 2NH_3$$

$$2NH_3 + H_2O + CO_2 \longrightarrow (NH_4)_2CO_3$$

$$(NH_4)_2CO_3 + CaSO_4 \longrightarrow (NH_4)_2SO_4 + CaCO_3$$

ii) Preparation of Urea: The main principle involved in the process of manufacture is combining pure ammonia with pure CO_2 and removing one molecule of H_2O from the resulting NH_4CO_3 to form Urea. The CO_2 and NH_3 are allowed to react in the liquid phase under greatly elevated pressure and temperature in presence of suitable catalysts and this process requires highly specialized equipment. The CO_2 and NH_3 are compressed and heated as they enter the converter where urea is formed. A large excess of NH_3 is used in order to increase the conversion rate.

Urea is manufactured by reacting anhydrous ammonia with CO_2 under higher pressure in presence of suitable catalyst. The intermediate unstable product ammonium carbamate is decomposed to urea:

SOIL FERTILITY AND NUTRIENT MANAGEMENT

$$N_2 + 3H_2 \longrightarrow 2NH_3$$
$$2NH_3 + CO_2 \longrightarrow NH_2COONH_4$$

30 atm (Ammo. Carbomate, Unstable intermediate product)

The unreacted NH_3 and CO_2 are removed by means of an evaporator still and are then recycled. The urea solution is pumped to the crystallizer where cooling and crystallization take place. The urea crystals are centrifuged and dried. This unstable intermediate product is decmposed and urea is recovered. The urea solution is then concentrated to 99 per cent and is sprayed into a chamber where urea crystals are formed.

$$NH_2COONH_4 \longrightarrow NH_2 CONH_2 + H_2O$$

(Ammo. Carbomate) (Urea, white crystalline substance)

During the preparation of urea, biuret is formed which is harmful. This biuret is formed when two molecules of urea are reacted eliminating NH_3.

$$NH_2.CO.N (H_2 + H.N) HCO.NH_2 = NH_2 CO. NH.CO NH_2 + NH_3$$

Urea   Urea   Biuret

In urea biuret should not be more than 1.5%.

Reactions of urea in soil

After application of urea in soil, it undergoes enzymatic hydrolysis mediated by Urease enzyme to produce an unstable compound designated as ammonium carbamate.

$$NH_2\text{-}CO\text{-}NH_2 + H_2O \xrightarrow{\text{Urease Enzyme}} NH_2COONH_4$$

(Urea) (Water)   (Ammo. Carbamate)

$$H_2COONH_4NH_3 + CO_2 \longrightarrow NH_3 + CO_2$$

(Ammo. Carbamate)   (Ammonia) (Carbon dioxide)

This NH_3 is converted to NH_4^+ ions by accepting one proton (H^+) from proton donor and subsequently forms of NH_4OH or any other NH compound depending upon the nature of the donor. Then after Ammonical-N (NH_4^+) form undergoes nitrification so as to produce nitrite and nitrates subsequently which is available for the plant growth.

Now a day's urea is used as fertilizer more compared to other nitrogenous fertilizers due to the following reasons:

a) Higher N content (44 to 46 per cent).

b) Good physical conditions.

c) Less acidic in residual effect compared to A/S.

d) Less cost per unit of N in production, storage and transport.

e) Lack of corrosiveness.

f) Suitable for foliar application, and

g) It is having of equal agronomical value compared to other nitrogenous fertilizers.

Slow release N fertilizers

Slow-release fertilizers are excellent alternatives to soluble fertilizers. Because nutrients are released at a slower rate throughout the season; plants are able to take up most of the nutrients without waste by leaching. A slow-release fertilizer is more convenient, since less frequent application is required. Fertilizer burn is not a problem with slow-release fertilizers even at high rates of application; however, it is still important to follow application recommendations. Slow-release fertilizers may be more expensive than soluble types, but their benefits outweigh their disadvantages.

Slow-release fertilizers are generally categorized into one of several groups based on the process by which the nutrients are released. Application rates vary with the different types and brands, with recommendations listed on the fertilizer label.

Pelletized

One type of slow-release fertilizer consists of relatively insoluble nutrients in pelletized form. As the pellet size is increased, the time it takes for the fertilizer to breakdown by microbial action is also increased.

Chemically Altered

A fertilizer may be chemically altered to render a portion of it water insoluble. For instance, urea is chemically modified to make Ureaform (ureaformaldehyde) — a fertilizer that is 38 percent nitrogen, 70 percent of which is water-insoluble. This percentage is often listed on fertilizer labels as the Percent W.I.N., or the percent of water-insoluble nitrogen. This form of nitrogen is released gradually by microbial activity in the soil. Because microbial activity is greatly affected by soil temperature, pH, aeration, and texture, these variables can affect the release of nitrogen from Ureaform. For example, there will be less fertilizer breakdown in acid soils with poor aeration — an environment unfavorable to soil microorganisms.

IBDU (isobutylidene diurea) is similar to Ureaform, but contains 32 percent nitrogen, 90percent of which is insoluble. However, IBDU is less dependent on microbial activity than Ureaform. Nitrogen is released when soil moisture is adequate. Breakdown is increased in acid soils.

Coated fertilizers

Controlled or slow-release fertilizers are broadly divided into uncoated and coated products. Uncoated products rely on inherent physical characteristics, such as low solubility, for their slow release. Coated products mostly consist of quick-release N-sources surrounded by a barrier that prevents the N from releasing rapidly into the environment.

Few examples of coated N fertilizers

Neem coated Urea is prepared by mixing Neem oil with Urea granules before application. As per the Fertilizer Control Order (1985) amendment, neem oil @ 0.350 kg is required to coat one ton of urea fertilizer.

Sulfur-coated urea (SCU), which is manufactured by coating hot urea with molten sulfur and sealing with polyethylene oil or a microcrystalline wax.

8.2. PHOSPHATIC FERTILIZERS

The phosphorus (P) nutrient of all Phosphatic fertilizers is expressed as P_2O_5. In soil, P is present as (i) Organic P and (ii) Inorganic P. The forms of inorganic P are $H_2PO_4^-$; HPO_4^{-2}; and PO_4^{-3}. Out of which, $H_2PO_4^-$ and HPO_4^{-2} ions are available to plant. In soil, water in is changed to HPO_4^{-2} and PO_4^{-3} ions with increase in pH.

$$H_2PO_4^- \xrightarrow{-H^+} HPO_4^{-2} \xrightarrow{-H^+} PO_4^{-3}$$

Firstly, the P in soil is immobile or slightly mobile around one cm diameter and therefore, they should be applied in root zone.

Secondly, the requirement of P is maximum in the initial crop growth stages. The crop takes up 2/3 of total P when the crop gains 1/3 of total dry matter and hence, the entire quantity should be applied at one time that is at the time of sowing as a basal dose.

Thirdly, water soluble-P is changed to insoluble form as Fe and Al $-PO_4$ (Phosphate) under acidic and calcium phosphate in calcareous or high Ca content or in higher pH soils and hence, there is no danger for the loss due to leaching and volatilization. The applied P remains as in available form in less quantity while greater quantity is changed to insoluble form.

Chemistry of P compounds

Phosphorus when burns gives P_2O_5 and with water, it forms HPO_3 (Metapohosphoric acid) and H_3PO_4 (orthophosphoric acid) $P_2O_5 + H_2O = 2\ HPO_3$; $HPO_3 + H_2O = H_3PO_4$. These H_3PO_4 is important in agriculture as it forms three compounds (salts) by replacing one hydrogen every time.

$$H_3PO_4 \xrightarrow{-H^+} H_2PO_4^- \xrightarrow{-H^+} HPO_4^{-2} \longrightarrow PO_4^{-3}$$

When H_3PO_4 combines with calcium, it forms three salts.

They are – i) $Ca(H_2PO_4)_2$ Monocalcium phosphate

ii) $CaHPO_4^-$ Dicalcium phosophate

iii) $Ca_3(PO_4)_2^-$ Tricalcium phosophate

Classification of Phosphatic fertilizers

The Phosphatic fertilizers are classified into three classes depending on the form in which H_3PO_4 combined with Ca.

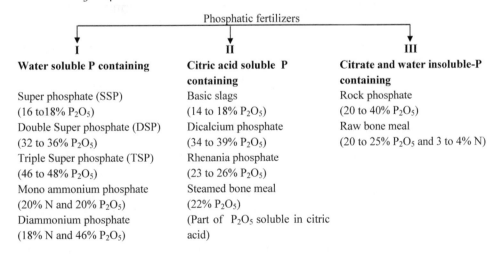

I	II	III
Water soluble P containing	**Citric acid soluble P containing**	**Citrate and water insoluble-P containing**
Super phosphate (SSP) (16 to18% P_2O_5)	Basic slags (14 to 18% P_2O_5)	Rock phosphate (20 to 40% P_2O_5)
Double Super phosphate (DSP) (32 to 36% P_2O_5)	Dicalcium phosphate (34 to 39% P_2O_5)	Raw bone meal (20 to 25% P_2O_5 and 3 to 4% N)
Triple Super phosphate (TSP) (46 to 48% P_2O_5)	Rhenania phosphate (23 to 26% P_2O_5)	
Mono ammonium phosphate (20% N and 20% P_2O_5)	Steamed bone meal (22% P_2O_5)	
Diammonium phosphate (18% N and 46% P_2O_5)	(Part of P_2O_5 soluble in citric acid)	

General Properties of Phosphatic Fertilizers

i) **Water soluble P containing fertilizer:**
   a) They contain water soluble-P as H_2PO_4 ion which can be absorbed quickly and available to plants when root system is not fully developed.
   b) Water soluble-P is rapidly transformed into water insoluble form in soil and hence there is no danger of loss due to leaching.
   c) These fertilizers should be used on slightly acidic, neutral to alkaline soils but not on acidic soils as the water soluble-P is changed to unavailable Fe and $Al-PO_4$.
   d) These fertilizers are applied when a crop requires quick start and for short duration crops.

ii) **Citric acid (1%) soluble P containing fertilizers:**
   a) They contain citrate soluble-P and hence this P is less available than water soluble-P.
   b) They are suitable for moderately acid soils because it gets converted into water soluble form. They are basic in reaction and Ca content.
   c) There are less chances of getting fixed by Fe and Al.
   d) They are suitable for long term crops and where immediate and quick start to crops is not important.

iii) **Citrate and water insoluble P fertilizers:**
   a) They are suitable for strongly acidic soils
   b) They contain insoluble P and hence not available to crops
   c) The P is available when ploughed with green manuring crop or organic residues.
   d) They are used for long duration crops and in large quantity 500 to 1000 kg/ha
   e) They are used where immediate effects are not important

Manufacturing of Phosphatic fertilizers

i) **Single super phosphate (SSP):** SSP is manufactured by mixing equal amounts of rock phosphate and concentrated H_2SO_4 (approximately 70%) and allowing to react for one minute in mechanical rotators. It is left for 12 hrs to harden and to cool down. It is then matured and after some weeks, it becomes ready for use. Due to free H_2SO_4 present in it, it is responsible for destroying gunny bags and hence first fill in polythene bags and then in gunny bags.

$Ca_3(PO_4)_2 + H_2SO_4 + 5H_2O = Ca(H_2PO_4)_2 \cdot H_2O + 2CaSO_4 \cdot 2H_2O$

It contains two part by weight Ca $(H_2PO_4)_2$ Monocalcium phosphate (16 to 18% water soluble-P) and three parts by weight gypsum. The formula of superphosphate is $Ca(H_2PO_4)_2 \cdot H_2O, CaSO_4 \cdot 2H_2O$. Superphosphate supplies P, Ca and S and due to gypsum, it improves physical conditions of soil when added to soil.

In double superphosphate, there is no separate process but gypsum is removed and when the P_2O_5 per cent of the content comes to 32 to 36, it is called double super phosphate.

In triple super phosphate, phosphoric acid is used instead of H_2SO_4 with calcium phosphate.

ii) **Diammonium phosphate: (DAP):** DAP or monoammonium phosphate is prepared by reacting phosphoric acid (H_3PO_4) with NH_3. In this if one H^+ ion of H_3PO_4 reacts with NH_3 it forms MAP ($NH_4H_2PO_4$) and NH_3 reacts with two H^+ ions of H_3PO_4 forms DAP [$(NH_4)_2 HPO_4$].

$$NH_3 + H_3PO_4 \longrightarrow NH_4H_2PO_4$$

Mono ammonium phosphate ($NH_4 H_2PO_4$)

$$2NH_3 + H_3PO_4 \longrightarrow (NH_4)_2HPO_4$$

DAP $(NH_4)_2 HPO_4$

8.3. POTASSIC FERTILIZERS

Potassium (K) is present in soil as:

i) Readily available forms as in soil solution and as exchangeable. These forms are available and plant absorbs these K forms as K^+ ion.

ii) Slowly available form as non-exchangeable i.e. fixed

iii) Relatively unavailable in the form of minerals (feldspars and micas etc.)

Firstly, the potash behaves partly like N and partly like P. From view point of the rate of absorption, it is required (absorbed) upto harvesting stage like N and like P, it becomes slowly available. Therefore, the entire quantity is applied at sowing time.

SOIL FERTILITY AND NUTRIENT MANAGEMENT

Secondly, potash being cation adsorbed on clay complex and hence leaching loss reduces. Leaching is greater in light soils than heavy textured soils. Therefore, like N, some time split application of K is desirable in sandy soil.

Thirdly, even though the soil contains enough potash or does not give response to crops, it becomes necessary to apply for the following reasons:

a) Maintaining K status of soil
b) For improving burning quality of tobacco
c) For neutralizing harmful effects of chloride in plant
d) For sugars or starch producing crops like potato, sweet potato, sugar cane, sugar beet, banana etc. for formation of sugar and starch.
e) For fibrous crops like sann, flex etc. to give strength to fibre
f) For the formation of pigments in crops like tomato, brinjal etc for quality purpose and it improves the luster and gives more colouration to the fruits of these crops by which more price can be fetched of the said products.

Classification of Potassic fertilizers

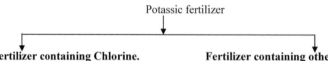

Fertilizer containing Chlorine.
Eg. KCl (Muriate of potash 58/60% K_2O). This is cheaper fertilizer and used extensively by cultivators for all crops except where chlorine is not desired in fertilizer

Fertilizer containing other than chlorine.
Eg. Sulphate of potash (K_2SO_4 48% K_2O)
Potassium Nitrate (KNO_3 44% K_2O, 13% N)
Sulphate of potash and magnesium (double salt of K and Mg, (Schoenite) K_2SO_4, $MgSO_4$ (25 to 30% K_2O)

Chemistry of K compounds

Potassium is not found in Free State in soil. As metal, it reacts with CO_2 forming K_2O and K_2O with H_2O gives KOH. For this reason, K in elemental form is not used as fertilizer. It must be combined with other element like chlorine or group of elements.

General Properties and manufacturing of Potassic fertilizers

1. **Muriate of Potash (KCl):** It is commonly marketed as a commercial fertilizer in granular form. However, it is also available in powder form. It is easily soluble in water. On application to the soil, it ionizes to dissociate into K^+ and Cl ions. K^+ like NH_4^+ gets attached or absorbed on the soil complex. As such, though muriate of potash is readily soluble in water, it is not leached.

The KCl is commercially manufactured using any one of the minerals; 1. Sylvinite or sylvite and 2. Carnallite.

There are two main steps involved in the manufacture of this fertilizer.

1. Mining of the K mineral

2. Separation of the main ingredient and purifying.

After purification, the desired sized Sylvinite /Carnallite mineral is dissolved in water to get a pulp and the reagent is added to the pulp. The reagent will form a coating or film round the NaCl molecules and this reagent added pulp is allowed to pass through a series of floatation cells in which air is introduced at the bottom in such a manner as to form a small bubble froth that attaches to NaCl. The commonly used reagents are (1) Sulphonated aliphatic alcohols of KCl and (2) 20% of mono-n-octadecyl- amines neutralized with acetic acid and a 0.5% solution of this is used for NaCl. Finally, the resultant product is Muriate of Potash.

It is found that at least 90% of the Potassic fertilizers consumed are in the form of muriate of potash. It contains 48 to 62% K_2O and 35 to 47% Cl. The commercial product is either granular or powdery having white or red colour.

2. **Potassium sulphate – K_2SO_4:** This is manufactured from kainite (KCl. Mg SO4, $3H_2O$) or langbeinite (K_2SO_4, Mg SO_4), the latter being commonly used. The raw materials required are the mineral langbeinite and KCl.

$$K_2SO_4 \cdot MgSO_4 + 2\ KCl \longrightarrow 2K_2SO_4 + MgCl_2.$$

The $MgCl_2$ is rejected. The KCl is used in the form of a mother liquor obtained from KCl manufacturing process. The method is found to be efficient only in the presence of a proper ratio between K_2SO_4 and $MgSO_4$ of the raw material and to maintain a proper ratio, KCl or Kainite is added. The reactions are found to take place in different ways.

Langbeinite, ground to pass through a 200 mesh sieve and the mother liquor from the KCl process is mixed and then the salts are recycled with water and fresh quantity of langbeinite. The reactions are allowed to take place in interconnected tank sand the crystallized K_2SO_4 is separated out by means of centrifuge. The used K_2SO_4 is separated out by means of centrifuge. The wet K_2SO_4 is dried and conveyed to storage place. It contains 48% K_2O. Only limited quantities of K_2SO_4 are manufactured, as the cost of manufacture is more, but is preferred for those crops where quality is desired or where chlorine toxicity will result when K Cl is used.

8.4. SECONDARY NUTRIENTS

Calcium, magnesium and sulphur -These three essential elements are called secondary nutrients because to the manufacturer of commercial fertilizers, these three nutrients are of secondary importance in the process of manufacture. However, these secondary nutrients are as essential as major nutrients.

A. Calcium

Calcium is absorbed by plant as Ca^{++} ion. It occurs in nature in carbonates, sulphides and hydroxide forms, complex calcium silicate and organic matter. It occur as exchangeable cation in neutral or slightly saline soils. Since the neutral and slightly alkaline soils of India are rich in Ca, there is no special need to apply materials to supply calcium.

Calcium is indirectly added to the soil through different fertilizers and soil amendments.

Approximate calcium content of some fertilizers and soil amendments

	Fertilizers/soil amendment	% of Ca^{++}
1.	Nitrogenous fertilizers	
	Calcium nitrate	19.5
	Calcium ammonium nitrate	8.1
	Calcium Cyanamide	39.1
2.	Phosphatic fertilizers	
	Single superphosphate	19.5
	Triple superphosphate	14.3
	Bonemeal	23.0
	Dicalcium phosphate	22.9
	Basic slug	33.9
3.	Soil amendments	
	Limestone	32.3
	Gypsum	29.2

Factors affecting calcium availability in soils

1. Total calcium supply.
2. Soil pH.
3. Cation exchange capacity (CEC).

CHEMICAL FERTILIZERS 147

4. Per cent (%) of Ca^{++} saturation on CEC.
5. Type of soil colloid.
6. Ratio of Ca^{++} to other cation in solution.

B. Magnesium

Magnesium is absorbed by plant as Mg^{++} ion. This essential plant nutrient has not been observed to be deficient on a wide scale in any part of India. As such no special material containing Mg has been recommended by the state Department of Agriculture.

Magnesium is however, indirectly applied to the soil through commercial fertilizers and soil amendments.

Approximate % of Mg in some fertilizers and soil amendments

	Fertilizers/soil amendment	% of Mg
1.	Nitrogenous fertilizers	
	Calcium ammonium nitrate	4.5
	Calcium nitrate	1.5
2.	Phosphatic fertilizers	
	Single superphosphate	0.3
	Basic slug	3.4
3.	Potassic fertilizers	
	Sulphate of potash	0.6
4.	Soil amendments	
	Dolomite limestone	4.0-10.6

Mg^{++} fertilizers

Dolomite is commonly applied to low Mg acid soils, $K_2SO_4 \cdot MgSO_4$ and $MgSO_4$ (Epsom salts) are the most widely used materials in dry fertilizer formulations. Other materials containing Mg are magnesia (MgO, 55 % Mg), magnesium nitrate [$Mg(NO_3)_2$-16 % Mg], magnesium chloride solution($MgCl_2 \cdot 10\ H_2O$, 8-9 % Mg)

$MgSO_4$, $MgCl_2$, $Mg(NO_3)_2$ and synthetic and natural Mg chelates are well suited for application in clear liquids and foliar sprays.

$K_2SO_4 \cdot MgSO_4$ (11% Mg) is the most widely used Mg additive in suspensions. A special suspension grade, 100 % passing through a 20-mesh screen of this material is available commercially.

C. Sulphur

Sulphur sources in soils

1. The main S-bearing minerals in rocks and soils are 1. Gypsum ($CaSO_4 \cdot 2H_2O$), 2. Epsomite ($MgSO_4 \cdot 7H_2O$), 3. Mirabilite ($Na_2SO_4 \cdot 10H_2O$), 4. Pyrite (FeS_2), 5. Sphalerite (ZnS), 6. Chalcopyrite ($CuFeS_2$).
2. Elemental S occurs in deposits over salt domes, in volcanic deposits and in deposits associated with calcite and gypsum.
3. Hydrogen sulphide (H_2S), an important commercial source of S, is a contaminant in many natural gas fields.
4. Organic compound containing S occurs in crude oil, coal and tar sand.
5. The original source of S was the metal sulphides in rocks.
6. Another source of S is the atmosphere. In regions where coal and other S-containing products are burned, SO_2 is released into the air and in later brought back to earth in precipitation.

Forms of S in soils

Sulphur is present in the soil in both organic and inorganic forms, although nearly 90 % of the total S in most non-calcareous surface soils exits in organic forms.

The inorganic forms are solution SO_4^{--}, adsorbed SO_4^{--}, insoluble SO_4^{--}, and reduced inorganic S compounds. Solution plus adsorbed SO_4^{--} represents the readily available fraction of S utilized by plants.

Sulphur fertilizers

Formally, FYM, ammonium sulphate, SSP a source of S were used and now their use is either restricted **or** they are replaced by other high analysis fertilizers which are devoid of S.

Therefore, S now become necessary to apply because of the following reasons…

1. Ammonium sulphate: a source of S is replaced by urea.
2. Sulphur containing SSP is replaced by DAP.
3. Higher use of KCL instead of K_2SO_4.

4. Decrease in use of FYM and
5. Use of high yielding varieties which absorb more quantity of nutrients.

A large amount of S is added every year indirectly through common fertilizers.

	Fertilizers/soil amendment	% S
1.	Nitrogenous fertilizers	
	Ammonium sulphate	23.7
	Ammonium sulphate nitrate	15.1
2.	Phosphatic fertilizers.	
	Single superphosphate	11-14
	Ammonium phosphate sulphate	15.4
3.	Potassic fertilizers	
	Potassium sulphate	17-18
	Potassium magnesium sulphate	22.7
4.	Soil amendments	
	Gypsum ($CaSO_4$)	23.5
	Gypsum ($CaSO_4 \cdot 2H_2O$)	18.6

8.5. MICRONUTRIENTS

Of the sixteen nutrients essential for plant growth, seven nutrients (Fe, Zn, Cu, Mn, B, Mo and Cl) are called micronutrients, because these nutrients are present in available forms in soil in very small quantities and the requirements by crops is also less.

Application of micronutrient fertilizers now becomes necessary as their deficiencies are observed in soil because of the following reasons:

1. Due to increase in irrigation facilities, the number of crops takes in a year increases.
2. Use of hybrid varieties which absorb more nutrients.
3. Intensive cultivation.
4. Reduction in use of organic manures like FYM, which supply these nutrients.
5. Use of high analysis fertilizers which are devoid of these nutrients.

The conditions mentioned above will definitely create the necessity of adding micronutrients either directly to the soil **or** through incorporating them in fertilizer mixture.

Out of Zn, Cu, Fe and Mn, Zn is considered as next to N in United States. The deficiency of Zn is also observed in Indian soils, similarly for Fe, but the deficiency of Cu and Mn are not observes on large scale in soil.

Viets (1967) has suggested the concepts of various pools of micronutrients.

1. Water soluble.
2. Exchangeable
3. Strongly adsorb, chelated or complex.
4. Secondary clay minerals and insoluble metal oxide.
5. Primary minerals.

A. Iron

Iron chlorosis is one the most difficult micronutrient deficiencies to correct in the field. In general, soil application of ionizable ferrous salt, such as ferrous sulphate, has not been satisfactory because of their rather rapid oxidation to much less soluble ferric iron.

Correction of iron deficiencies is done mainly with foliar spray to alleviate mild chlorosis. However, several sprays 7 to 14 days apart may be needed to remedy more severe iron deficiencies. Inclusions of several drops per liter of mild detergent to serve as a wetting agent are normally necessary.

Factors affecting the availability of iron in soils and plants.

1. Soil pH
2. The effect of bicarbonate, phosphate, calcium and potassium ions.
3. Organic matter.
4. Soil texture.
5. Water logging or submergence.

Wallace and Hunt (1960) considered the following causative factors responsible for iron chlorosis in horticultural plants.

1. Low Fe supply.
2. High $CaCO_3$.
3. HCO_3 ions in soil or water.
4. Over irrigation or high water condition.
5. High phosphate.
6. High level of Mn, Cu and Zn.

7. Lower or higher temperature.
8. High light intensity.
9. High levels of NO_3^- nitrogen.
10. Unbalance cation ratios.
11. Poor aeration.
12. Root damage by nematodes or other organisms.
13. Viruses.
14. Certain organic addition to the soil.

Some sources of fertilizer iron

Ferrous sulphate	$FeSO_4 \cdot 7H_2O$	19
Ferric sulphate	$Fe_2(SO_4) \cdot 4H_2O$	23
Ferrous oxide	FeO	77
Ferric oxide	Fe_2O_3	69
Ferrous ammonium sulphate	$(NH_4)_2SO_4 \cdot FeSO_4 \cdot 6H_2O$	14
Ferrous ammonium phosphate	$Fe(NH_4)PO_4 \cdot H_2O$	29
Iron chelates	NaFeEDTA	5-14
	NaFeHEDTA	5-9
	NaFeEDDHA	6
	NaFeDTPA	10

C. Zinc

Zinc sulphate is the principle zinc salt used in fertilizers. It is a white, water soluble salt. The salt is also suitable for foliar spray. Other elements used in sulphate form are Mn and Fe. The relative insoluble Zn sources like zinc oxides, zinc carbonate and zinc phosphate compare well with zinc sulphate and these can serve as alternative zinc sources.

Soil conditions most often associated with zinc deficiencies are:

1. Acid sandy soil low in total zinc.
2. Neutral or basic soils, especially calcareous soils.
3. Soils with high content of fine clay and silt.
4. Soils high in available phosphorus.
5. Some organic soils and
6. Sub-soils exposed by land leveling operation or by wind and water erosion.

152 SOIL FERTILITY AND NUTRIENT MANAGEMENT

Factors affecting the availability of zinc in soil.
1. Soil PH.
2. Organic matter.
3. $CaCO_3$.
4. Clay minerals.
5. Phosphates.

Some sources of fertilizers zinc.

Zinc sulphate monohydrate	$ZnSO_4.H_2O$	35
Zinc sulphate heptahydrate	$ZnSO_4.7H_2O$	23
Zinc oxide	ZnO	78
Zinc carbonate	$ZnCO_3$	52
Zinc sulphide	ZnS	67
Zinc phosphate	$Zn_3(PO_4)_2$	51
Zinc chelates	$Na_2ZnEDTA$	14
	Na Zn NTA	13
	Na Zn HEDTA	9

C. Manganese

Manganese sulphate is widely used for correction of manganese deficiency. It may be applied to soil or directly to crop as a foliar spray. In addition to simple inorganic compounds, fertilizer Mn is commercially available in chelated, organically complexed and fritted forms. The natural organic complexes and chelated sources of Mn are best suited for spray applications.

Factors affecting the availability of manganese in soil:
1. Soil PH
2. Organic matter
3. Effect of $CaCO_3$
4. Effect of water logging.
5. Effect of size of particles.
6. Fe/Mn ratios.
7. Effect of clay minerals.
8. Effect of cropping and soil management practices.
9. Effect of air drying.

10. Effect of microorganisms on soil manganese transformation.
11. Effect of fertilizers and soil amendments on Mn availability.
12. Liming.
13. Effect of Na and other exchangeable cations.
14. Green manuring.
15. Manganese – iron interrelationship.

Sources of Mn used for fertilizer.

Manganese sulphate	$MnSO_4.4H_2O$	16-28
Manganous oxide	MnO	41-68
Manganese carbonate	$MnCO_3$	31
Manganese chloride	$MnCl_2$	17
Manganese oxide	MnO	63
Synthetic chelate	MnEDTA	5-12

D. Copper

The usual copper source is copper sulphate ($CuSO_4.5H_2O$). This salt is quite soluble in water and is comparable with most fertilizer materials. Soil and foliar application are both effective, but soil application are more common.

Factors affecting the availability of copper in soil:

1. Soil PH
2. Calcium carbonate.
3. The organic matter or humus.
4. Microbial activities.
5. Clay minerals and other inorganic materials.
6. Fertilizer practices such as application of phosphates.

Copper compound used as fertilizers.

Copper sulphate (chalcanthite)	$CuSO_4.5H_2O$ $CuSO_4.H_2O$	25
(monohydrate)		35
Copper nitrate	$Cu(NO_3)_2.3H_2O$	
Copper ammonium phosphate	$Cu(NH_4)PO_4.H_2O$	32
Copper chelates	Na_2Cu EDTA	13
	NaCu HEDTA	9

SOIL FERTILITY AND NUTRIENT MANAGEMENT

E. Boron

Boron is only non-metal among the micronutrient elements. It has a constant valency of +3 and has a very small ionic radius. It is absorb by the plant as HB_4O_7 (bimorate) and BO_3^- cborate.

Boron is found deficient in calcareous soil as it is changed to calcium borate which is insoluble and hence it is applied as fertilizers

Soil where boron deficiency is commonest

1. Soils low in boron, such as those derived from acid igneous rocks or from fresh water sedimentary deposit.
2. Highly leached acid soils. Laterites and fall in this group.
3. Light textured coarse sandy soils.
4. Soil low in organic matter.
5. Alkaline soils with a high amount of free $CaCO_3$.

Factors affecting the availability of boron in soil

1. Parent material
2. Texture
3. $CaCO_3$
4. Effect of cultivation
5. Irrigation water

Principle Boron fertilizer

Borax	$Na_2B_4O_7.10H_2O$	11
Boric acid	H_3bO_3	17
Sodium tetraborate	$Na_2B_4O_7.5H_2O$	14-15
Fertilizer borate-48(Agribor, Thonabor)		
Fertilizer borate	$Na_2B_4O_7$	21
Solubor	$Na_2B_4O_7.5H_2O+Na_2B_{10}O_{16}.10H_2O$	20-25

Borax or sodium borate is white salt soluble in water. As such it is suitable for soil application and foliar spray.

F. Molybdenum

Molybdenum, which occurs in earth's crust and in soil in soils in extremely small quantities, is usually found in concentration of less than 1 ppm in plants. Molybdenum is required in very small quantity and also present in sufficient quantity in some seeds and soils and hence generally its fertilizer are not used.

Factor affecting the availability of molybdenum in soils.
1. Soil pH
2. Organic matter
3. Calcium carbonate
4. Texture
5. Phosphate and sulphur fertilizers
6. Manganese
7. Nitrogenous and Potassic fertilizers
8. Oxides of iron and aluminium

Sources of molybdenum used for fertilizer

Ammonium molybdate	$(NH_4)_6Mo_7O_{24}.2H_2O$	54
Sodium molybdate	$Na_2MoO_4.2H_2O$	39
Molybdenum trioxide	MoO_3	66

G. Chlorine

Although, chlorine is required by plants in large amount, its deficiency has not been reported so far in crop production. The reason is that air and water contain sufficient amount of chlorine which can be readily utilized by the plants. It is estimated that approximately five pound or more of chlorine are added to the soil annually through rain water. Besides, some common fertilizer, like ammonium chloride (69-70 % Cl⁻) and KCl (47.3 % Cl) supply indirectly chlorine to the soil in readily available form for crop use.

Out of seven micronutrients B, Mo and Cl are present in the form of anions and absorb by the plant as anions. Rest of four micronutrients (Fe, Zn, Cu and Mn) is generally applied both as soil and foliar application as their sulphates at the time of deficiency.

Now the day micronutrients carriers which contain all micronutrient are available in the market. It is not advisable to use such types of materials as the nutrients which are sufficient in soil will reduce the availability of other nutrients based on soil test value should only be applied.

Role of chelates in micronutrient availability

When the commonly used compounds containing micronutrients applied to soil, they get fixed in the soil in such a way that micronutrients become unavailable to growing plants. As the materials containing the micronutrients are usually applied in small quantity, their beneficial effect on the crop become still limited. To increase the availability of micronutrients and make them slowly available over a long period, chelated compounds are formed.

What is chelate?

Chelates is a compound in in which metallic cations (Ca^{++}, Mg^{++}, Fe^{++}, Zn^{++}, Cu^{++} and Mn^{++}) is bound to an organic molecules. In chelated form, these cations are protected from reactions with inorganic constituents that would make them unavailable for uptake.

Terms such as chelates, chelated complex or chelated compounds refers to organo-metallic molecules of varying sizes and shapes in which the organic part bind the metal cation (nutrient) in a ring-like structure. The organic redial (chelating agent) is known as "Ligand". (Ex. EDTA) which is negatively charged and form covalent bond with positively charged metal cation. The nutrients which can be chelated are Ca, Mg, Fe, Zn, Cu, and Mn. Such compounds usually do not react with soil complex and remain in solution at a much higher pH. Thus, when commonly used materials containing micronutrients become unavailable to the plants at a high pH, the chelated compounds supply nutrients to them.

In absence of chelates, when an inorganic iron salt such as ferric sulphate is added to calcareous soil, most of iron is quickly rendered unavailable by reaction with hydroxide.

$$Fe^{+3} + 3\ OH \longrightarrow FeOOH + H_2O$$

If iron is added as its chelates (Fe-EDTA), the Fe remains in chelate form, which Fe is available for uptake.

The mechanism by which micronutrients from chelates are absorb by the plant is obscure. The chelates keep the metal available in soil and increase transformation of metals once they are absorb by the plant.

Stability of chelates: They vary in their stability. Fe chelates tends to more stable than those of Zn and Cu, which is turn are more stable than Mn. Thus, when Zn chelated is added to soil, Fe will replace the Zn.

$$\text{Zn chelates} + Fe^{++} \longrightarrow \text{Fe chelates} + Zn^{++}$$

Since, Fe chelate is more stable than the Zn, counterpart, the reaction goes to right side and Zn is subjected to reaction with soil.

The common chelating agents are:

- **EDTA:** Ethylene Diamine Tetra Acetic Acid.
- **DTPA:** Diethylene Triamine Penta Acetic Acid.
- **CDTA:** Cyclohexa Diamine Tetra Acetic Acid.
- **HEDTA:** Hydroxy Ethylethylene Diamine Triacetic Acid.
- **EDDHA:** Ethylene Diamine Dico-hydroxyphenyl Acetic Acid.
- **NTA:** Nitrico Triacetic Acid.
- **EGTA:** Ethylene Glycol-big Tetra Acetic Acid.

For four micronutrients, the following is the order of stability of some major chelates:

- **Fe:** EDDHA>DTPA>CDTA>EDTA>EGTA>HEDTA>NTA
- **Zn:** DTPA>CDTA>HEDTA>EDTA>NTA>EGTA
- **Cu:** DTPA>HEDTA>CDTA>EDTA>EDDHA>EGTA>NTA
- **Mn:** DTPA>CDTA>EDTA>EGTA>HEDTA>NTA

Application of micronutrients:

1. Chlorine is not applied as fertilizer because it is indirectly applied through irrigation water.
2. Molybdenum is also not applied as fertilizer because it is present in seeds and required in very small quantity.
3. Boron is found to be deficient in calcareous soils because it is changed to calcium borate which is insoluble and hence it is applied as fertilizer.
4. Zn, Fe, Cu and Mn are generally applied as soil and foliar application as their sulphates. In fact, only the deficient micronutrient from soil test value should be applied.

8.6. MIXED FERTILIZERS OR FERTILIZER MIXTURES:

General: For application of individual nutrient, fertilizers of the nutrients are also applied separately which increased labour, storage, transport and application cost. In order to avoid the said difficulty, the fertilizer mixtures are used.

Fertilizer mixtures: A mixture of two or more straight fertilizer materials is referred to as fertilizer mixture (mixed fertilizer). Sometimes, complex fertilizers containing two plant nutrients are also used in formulating fertilizer mixtures.

Fertilizer mixtures are formulated products made by mixing together two or more fertilizer materials. In all the state of India, fertilizer mixture containing two or three major nutrients has been introduced.

Types of fertilizer mixtures

I. **Open formula fertilizer mixtures**: The ingredients mixed in this type of fertilizer mixture in terms of kind and quality is disclosed by the manufacturer. This will helpful for the cultivators to know the ingredients of fertilizer for the application of same in the particular crop in suitable amounts.

II. **Closed formula fertilizer mixtures**: The ingredients or straight fertilizer used in this fertilizer mixture are not disclosed by the manufacturer. It is called as a trade secret of the industry. So it is not possible for farmers to know the type and quantity of ingredients used in this fertilizer mixture. Thus, farmers can't choose a correct mixture for their used in production of crops.

Materials used in fertilizer mixture

1) **Supplier of plant nutrients:** The type of grade of fertilizer mixture to be prepared should be decided, the straight fertilizers are chosen accordingly to compatibility in mixtures. The quantity of each fertilizer is calculated for preparation of desired quantity of preparing fertilizer mixture. Thus these are the primary materials most essential in preparing mixed fertilizers.
   - N is supplied mainly through, AS urea, CAN, ASN, Oilcakes and sludge.
   - P_2O_5 is supplied mainly through SSP and Bone meal.
   - K_2O supplied mainly through KCl and K_2SO_4.
   - Complex fertilizers are also used in the production of fertilizer mixtures.

2) **Conditioners:** To prepare mixed fertilizer in good drilling condition and to reduce caking, low grade organic materials are usually added at the rate of about 45 kg/ton. These materials known as conditioners which are tobacco stem, peat, groundnut hulls and paddy hulls.

3) **Neutralizer of residual acidity:** If the nitrogen fertilizers used are acidic in nature like as, urea, basic materials like dolomite, limestone is added to counteract the acidity, such materials are known as neutralizer of residual acidity.

4) **Filler:** Filler is the make-weight material added to a fertilizer mixture. It is added to make up the difference between the weight of the added fertilizer required to supply the plant nutrients and the desired quantity of fertilizer mixture. The common filler materials used are sand, soil, ground coal ashes and various other waste products.

Advantages of fertilizer mixtures

1. Less labour is required to supply a mixture than to apply its various components separately.
2. Use of fertilizer mixtures leads to balanced manuring.
3. The residual acidity of fertilizer can be effectively controlled by the use of proper quantity of liming materials in the mixtures.
4. Micronutrients which are applied in small amounts to soil can be incorporated in fertilizer mixtures. This facilitates uniform soil application of plant nutrients required in small quantities
5. Mixtures have the better physical condition and are more easily applied than many straight fertilizers. This advantage is most marked when mixtures are prepared in a granular form.
6. Fertilizer mixture containing suitable filler improves the physical condition of soil.

Disadvantages of fertilizer mixtures

1. Their use does not permit application of individual nutrient which may best suit the need of a crop at specific times.
2. The unit cost of plant nutrients in mixture is usually higher than those of straight fertilizers.
3. Farmers often used mixtures without careful study of their need, thus applying too little of some nutrients and much more of other.
4. Fertilizer mixture of a particular grade suitable for a particular crop cannot be applied profitable to all crops.

Fertilizer grade: The fertilizer grade refers to minimum guarantee of plant nutrients in terms of total N, available P_2O_5 and water soluble K_2O. A 5:10:10 grade fertilizer mixture is guaranteed 5 % total N, 10 % available P_2O_5 and 10 % water soluble K_2O.

Incompatibility in Fertilizer mixture

The preparation of satisfactory mixed fertilizers requires detail knowledge of the chemical and physical characteristic of the individual material, their behavior in mixture under different condition of storage, transportation as well as under varying circumstances of atmospheric temperature and humidity.

In mixing straight fertilizers, following general rules should be observed.

1) Fertilizer containing ammonia (AS., NH_4NO_3) should not be mixed with basically reactive fertilizers (lime, basic slag, rock phosphate, $CaCN_2$) as loss of N may result though escape of gaseous ammonia.

2) All water soluble phosphatic fertilizers (SSP, TSP, ammonium phosphate) should not be mixed with those fertilizers that contain free lime because this convert a portion of the soluble P into insoluble form.

3) Easily soluble and hygroscopic fertilizers like CAN, urea and potash salts) tend to cake or form lumps after mixing. Such fertilizers should be mixed shortly before use.

4) Slightly acidic SSP may damages gunny bags and drilling implements due to free H_2SO_4.

8.7. COMPLEX FERTILIZERS

General: Due to uneconomical and labour cost of using individual fertilizer, the fertilizer mixture was prepared and they were used. These fertilizer mixtures were not homogenous, containing less quantity of N, P, K and many times inferior quality of materials were used. Due to these difficulties and as manufacturing equipments were made available, complex fertilizers have been prepared. These complex fertilizers contain the nutrients of grade mentioned, homogenous, granular and good physical conditions.

Complex fertilizers: Commercial complex fertilizers are those fertilizers which contain at least two or three of the primary essential nutrients. It is also known as multiple-nutrient fertilizers.

Incomplete complex fertilizers: When complex fertilizers contain only two of the primary nutrients, it is designated as **incomplete** complex fertilizers. eg. DAP (18 % N, 46 % P_2O_5).

Complete complex fertilizers: When complex fertilizers contain all the primary essential nutrients (N, P_2O_5 and K_2O are designated as complete complex fertilizers eg. IFFCO'S NPK (12:32:16, 10-26-26).

At present complex fertilizers obtained by chemical reaction are considerably more important than those obtained by mechanical mixing or fertilizer mixtures.

Characteristics of complex fertilizers

1. They usually have a high content of plant nutrient more than 30 kg per 100 kg of fertilizer. As such they are also called "High Analysis Fertilizer".
2. They usually have a uniform grain size and good physical condition.
3. They supply N and P to the soil in available form in one operation. Nitrogen is present in NO_3^- or NH_4^+ form and P is present in water soluble form up-to 50-90 % of total P_2O_5.
4. They are cheaper as compared to individual fertilizer on the basis of per kg of plant nutrient.
5. Transport and distribution cost is reduced on the basis of per kg of plant nutrients.
6. They are non-caking and non-hygroscopic, thus safer for storage.

Advantage of complex fertilizers

1. The complex fertilizers contain more nutrients. So, its application is advantages in comparison to straight fertilizer, which is need to be applied separately.
2. The nutrients remain in combination in complex fertilizers and they do not separate in any condition.
3. Less cost is involved in transportation and application of fertilizer.
4. They are available in different grades, according to need of the soils and crops.
5. Provides opportunity to apply two or three plant nutrients in single operation.
6. Being granular, it is easy to apply by broadcasting contains micronutrients in addition to primary nutrients.

Disadvantages of Complex fertilizers:

1. Complex fertilizers may not always supply balanced nutrients to the crop.
2. Contain NPK in certain proportion **or** grades, but the crop may be needed NPK in quite different proportion.
3. If the complex fertilizers do not meet the need of the crop, straight fertilizer needs to apply for ensuring a balanced fertilization.

SOIL FERTILITY AND NUTRIENT MANAGEMENT

Considering the incompatibility, the chart is given below which can be used while preparing fertilizer mixture.

	1	2	3	4	5	6	7	8	9	10	11	
1	✓	✓	X	X	*	✓	✓	✓	X	✓	✓	Muriate of Potash
2	✓	✓	X	X	*	✓	✓	X	X	✓	✓	Sulphate of potash
3	X	X	✓	✓	*	X	*	*	✓	✓	✓	Sulphate of ammonia
4	X	X	✓	✓	*	X	*	*	✓	✓	✓	Calcium amm. nitrate
5	*	*	*	*	*	*	*	*	*	*	*	Sodium nitrate
6	✓	✓	X	X	*	✓	✓	X	X	X	*	Calcium cynamide
7	✓	✓	*	*	*	✓	✓	✓	✓	*	*	Urea
8	✓	X	*	*	*	X	✓	✓	✓	*	*	Superphosphate single or triple
9	X	X	✓	✓	*	X	✓	✓	✓	✓	✓	Ammon. Phosphate
10	✓	✓	✓	✓	*	X	*	*	✓	✓	✓	Basic slag
11	✓	✓	✓	✓	*	*	*	*	✓	✓	✓	Calcium carbonate

Guide for mixing fertilizers

✓	Fertilizer which can be mixed
*	Fertilizer which may be mixed shortly before use
X	Fertilizer which can not be mixed

CHEMICAL FERTILIZERS

8.8. LIQUID FERTILIZERS

For intensive high yield and quality crop production, liquid fertilizers are preferred. This helps when both water and soluble fertilizers are delivered to crops simultaneously through a Drip Irrigation System ensuring complete plant nutrition such as N, P, K, Ca, Mg, S & Micronutrients which are directed to the active root zone in well balanced proportion.

Advantages of liquid fertilizer

- Nutrient availability to the plant is improved
- Nutrient uptake efficiency is increased
- Fertilizer Application rates & Water Requirements are reduced
- Losses by Leaching are minimized
- Salt Injuries & damages to Root & Foliage are prevented
- Soil Compaction is reduced due to less field operations
- Weed population is decreased

Liquid fertilizers and their solubility

Sr. no.	Fertilizer	Grade (N-P-K)	Solubility (g/lit) at 20°C
Water soluble special fertilizers			
1.	Mono Ammonium Phosphate (MAP)	12-61.0	282
2.	Mono Potassium Phosphate (MKP)	0-52-34	230
3.	potassium Nitrate (Multi-K)	13-0-46	316
4.	Sulphate of Potash	0-0-50	111
5.	Ortho Phosphoric Acid	0-52-0	457
Conventional WS Fertilizers			
1	Urea	46-0-0	1100
2	Potassium Chloride (Red)	0-0-60	347
3	Potassium sulphate (White)	0-0-50	110
4	Ammonium Sulphate	21-0-0	760

8.9. NANO-FERTILIZERS

Nanotechnology has progressively moved away from the experimental into the practical areas, like the development of slow/controlled release fertilizers, conditional release of pesticides and herbicides, on the basis of nanotechnology has become critically important for promoting the development of environment friendly and sustainable agriculture.

Indeed, nanotechnology has provided the feasibility of exploiting nanoscale or nanostructure materials as fertilizer carriers or controlled release vectors for building of so-called "smart fertilizer" as new facilities to enhance nutrient use efficiency and reduce costs of environmental protection.

Encapsulation of fertilizers within a nanoparticle is one of these new facilities which are done in three ways

a) The nutrient can be encapsulated inside nonporous materials,
b) Coated with thin polymer film and
c) Delivered as particle or emulsions of nanoscale dimensions.

In addition, nano fertilizers will combine nano devices in order to synchronize the release of fertilizer-N and -P with their uptake by crops, so preventing undesirable nutrient losses to soil, water and air via direct internalization by crops, and avoiding the interaction of nutrients with soil, microorganisms, water, and air. Among the latest line of technological innovations, nanotechnology occupies a prominent position in transforming agriculture and food production.

Some of the major evident benefits of nano fertilizer are as under:

- The quantity required for nano fertilizer application is considerably reduced as compared to conventional fertilizers.
- Nano fertilizer will help to boost the crop production efficiently besides reducing nutrient losses into the surrounding water bodies (Eutrophication).
- Nano-structured formulation might increase fertilizer efficiency and uptake ratio of the soil nutrients in crop production, and save fertilizer resource.
- Nano-structured formulation can reduce loss rate of fertilizer nutrients into soil by leaching and/or leaking.

8.10. FERTILIZER USE EFFICIENCY:

The increase in yield of the harvested fraction of the crop per unit of nutrient supplied by fertilizer is called Fertilizer use efficiency. Fertilizers are applied to supplement nutrient requirement of the crop. It should not be looked as a substitute to organic sources. After determination of nutrient requirement of a crop for a given yield and contribution of nutrients from different sources, particularly, from the soil source, it is necessary to supplement the balance from the inorganic sources. These are determined by field experimentation supplemented by pot-culture, laboratory and green house studies, if necessary. When a fertilizer is applied all of its nutrient(s) are not absorbed by the crop. The interactions between soil-crop-season and other factors are quite significant. Only a fraction of the nutrient(s) is utilized by the crop. Efficiency in any system is an expression of obtainable output with the addition of unit amount of input. The ratio of energy intake and energy of the produced biomass i.e. of input and output is called ecological efficiency. This can be studied at any trophic level. Fertilizer use efficiency is the output of any crop per unit of the nutrient applied under a specified set of soil and climatic conditions.

Techniques of increasing fertilizer use efficiency

To increase the fertilizer use efficiency the nutrient must be available at the time of its requirement by the crop, in right form and quantity. On application there occur certain inevitable/evitable losses of nutrients that reduce the efficiency. The losses are due to: (i) leaching, (ii) volatilization, (iii) immobilization, (iv) chemical reaction between various components in the mixture, (v) change in capacity to supply nutrients, and (vi) unfavourable effects associated with fertilizer application. Each component of loss can be reduced to a great extent by management of the soil fertilizer crop system. This requires knowledge and experience on (i) how much of the fertilizer to be applied, (ii) what/which (type of fertilizer) to be applied, (iii) when to be applied (time of application), (iv) how (method of application), (v) where (placement of fertilizer) and (vi) other considerations (cost, availability of fertilizer, labour, ease of application, awareness on benefits of fertilizer use, etc.).

(i) *How Much*

Inorganic source is a supplement to other sources of nutrients. Among other sources, the most important one is soil source. Availability of nutrients from soil and fertilizer sources can be estimated from field experiments involving response to fertilizers and tracer techniques (using radio-active isotopes).

(ii) *Type of fertilizer*

Fertilizers vary with respect to their solubility besides their grade. Choice of fertilizer is location specific and needs to be found out by field experimentation. The choice is more with respect to nitrogen and phosphatic fertilizers than for potassic. Studies on crop response is also more for N than for P or K fertilizers because leaching loss is more in nitrogenous fertilizer and its residual effect is nil or negligible. In case of P, its indirect, residual and cumulative effects are more important. Nitrogen in form of NO_{3-} is subject to more leaching. Leaching loss is also more in wet than in summer and in sandy soils than in clayey soils. Losses can be minimized by choosing suitable time and method of application.

(iii) *When to apply*

It necessarily means time of application. The objective of time of application is to get maximum benefit from the fertilizer nutrient. If the nutrient is applied too earlier than the time of requirement, it is lost in different ways or is absorbed more than required. If applied late it is either not absorbed or if absorbed not utilized for the purpose and only gets accumulated in plant parts. Some amendments need to be applied before commencement of crop season so that it reacts well with the soil and becomes available to the crop after sowing/planting.

(iv) *Where to apply*

The objective of placement of fertilizer is to make the nutrient available easily to the crop. It should be near to the roots. Application may be surface broadcast, at furrow bottom, placed deep at or slightly below the root zone, top dressed, side dressed or to foliage. This depends on type of crop, rooting pattern, feeding area and ease of application. The choice of method of application depends on soil-crop-fertilizer interaction too.

Factors Affecting Fertilizer Use Efficiency

The object of manuring is to improve the nutritional status of the soil by increasing the store of nutrients present and thus to raise the yield from a lower to a higher level. Consequently, the response of a particular crop to a given fertilizer application cannot be foretold because a material so readily subjected to change is placed in contact with the soil and the crop, which react with it both chemically and biologically, and thus affect its efficiency.

Again there is the soil moisture which has a tremendous effect upon the crop, and upon the fertilizers. If there is an excess or deficiency of moisture, the full efficiency of fertilizer cannot be expected. In the event of such a complex situation and attempt has been made to highlight the various factors affecting the response of crop to fertilizer and to suggest the soil and crop management to increase the efficiency of fertilizer.

A. **Crop Characteristics:**
   1. Kind of plant and root system

CHEMICAL FERTILIZERS

2. Varieties
3. Plant populations
4. Rotation and crop residues

B. Soil Characteristics:
1. Nutrient status of the soil
2. Soil moisture
3. Soil reaction
4. Soil temperature
5. Physical condition of the soil
6. Chemical nature of the soil
7. Effect of soil amendment

C. Fertilizer Characteristics and Fertilizer Manipulations:
1. Types of fertilizers.
2. Time of application
3. Method of application
4. Use of nitrate inhibitors.
5. Use of chelating substances.

A. Crop Characteristics

1. **Kind of plant and its root-system:** The response to the application of fertilizers may differ widely from crop to crop because of a number of factors among which the yield level and root characteristics are the most important. The yield level has a direct effect on the amount of nutrient removed from the soil. Although the nutrient uptake by different crop is a fairly good guide for the response of crops to fertilizers, the root characteristics of different plant species might often modify the response to different fertilizers. The root system of different plant species vary greatly in rapidity and extent of development. The extensive root system of corn exploited the soil more thoroughly than the limited root system of potatoes and consequently, the former used more nutrients from the soil and less from the fertilizers, whereas the reverse was true in case of potatoes. Thus the response to fertilizer will be more in case of potatoes than in case of corn.

The cation exchange capacity of the roots of dicotyledonous plants is much higher than that of monocotyledonous plant and absorbs more divalent cations such as calcium and less of monovalent cations such as potassium. In contrast, plants with a low root cation exchange capacity absorb less than of the divalent and more of the monovalent cations. Thus the response to K may be higher in the case of monocotyledonous plants than that in the case of dicotyledonous plants.

2. **Varieties:** Marked differences in response to fertilizers have been obtained within varieties of the same crop. In many cases, these differences may be due to differences in plant types. Some of the highly responsive varieties of cereal crops have got dark-green upright leaves which do not shade the lower leaves, intercept the maximum amount of sunshine and thus, make the most efficient use of added fertilizers.

3. **Plant population:** Sufficient plant population is one of the most important factors affecting response to the fertilizer application. In an experiment at Dharwad, with the lowest plant populations of sorghum (91,000), the highest response to the application of nitrogen was obtained at 150 kg N/ha, but with medium (136,000) and high (272,000) plant population, the maximum response was obtained with 200 kg N/ha.

4. **Rotation and Crop Residues:** The crop rotation has a profound effect on the fertilizer use efficiency and fertilizer requirements of the crops in the rotation. The legumes affect the nutrient status of the soil for the succeeding crop differently from exhaustive crops like jowar and maize. The crop requiting high levels of fertilizers, such as potato or hybrid maize, may not use the fertilizer applied to them fully and thus some quantities of nutrient element may be left in the soil and they may be available for the succeeding crop. On the other hand, if sub-optimal doses of fertilizers are applied to a crop, they may leave the soil in a much exhausted condition and the fertilizer requirement of the succeeding crop may be increase. The legumes leave N-rich root residues in the soil for the succeeding crops and thus, reduce its N-requirement.

B. Soil Characteristics

1. **Nutrient Status of the soil:** The response of crop to fertilizer application directly depends on chemical composition of the soil in respect to the available plant nutrient. On the basis of soil testing, the soils are related as 'low', 'medium' or 'high' in plant nutrients and suitable fertilizer doses are recommended. A 'low' rating in phosphorus means that crop in such soils should respond very readily to phosphate application. If the rating is 'medium' the response is probable: and if the rating is 'high' there may be little or no response to the applications of the phosphorus fertilizer.

2. **Soil moisture:** In most cases under irrigated conditions, fertilizers have greatly helped to increase crop yields by having favorable effects on the mass and distribution of roots. Nutrient absorption is affected directly by the level

of soil moisture as well as indirectly by the effect of water on the metabolic activities of the plant, soil aeration and the concentration of the soil solution.

3. **Soil reaction:** Soils differ markedly in their reaction of pH. The soil microorganisms respond very markedly to soil reaction which has direct and indirect effect on crop growth. Since the soil reaction has a profound influence on the availability of plant nutrients, there is definitely likely to be a different response to the application of fertilizers on soil differing in their reaction. The fertilizer practices have, therefore to be greatly modified for soil reactions. Another possibility may be to modify the soil reaction so as to make it most favorable to crop growth and fertilizer efficiency.

4. **Soil temperature:** Low soil temperatures, become a limiting factor in seed germination. The power of root cells to cumulate various nutrient ions within a certain temperature range is related directly to temperature. The nitrate production in soil increases with temperature up to about $30\ ^0C$ and then decreases with a further increase in temperature.

5. **Physical condition of the soil:** The physical condition of the soil largely determines the way in which it can be utilized by the plants. It is a resultant of the size, shape, arrangement and mineral composition of the soil particles as well as of the volume and form of its pores.

6. **Chemical properties of the soil:** Some fertilizer materials leaves acid residues in the soil, other a basic residue and still others might have no influence on the soil reaction. Fertilizer carriers of P and K have generally little influence on soil acidity, whereas the carriers of nitrogen have a considerable effect on soil pH.

7. **Effect of soil amendments:** Lime is the commonest material used for correcting the acidity of the soil. Lime affects plant growth mainly by correcting the soil pH and, thus, making available a large number of plant nutrients. Thus on acid soils, even responses to fertilizers are very much reduced without the use of lime. Likewise, gypsum and pressmud, like soil amendments, produce a beneficial effect in alkali soils in improving the efficiency of fertilizers.

C. Fertilizer Characteristics and Fertilizer Manipulation

1. **Types of fertilizers:** Fertilizers may be differing from one another in different ways. They may differ in their nutrient contents or in the form of nutrients. Thus in the case of N-fertilizers, the nutrients may be supplied in the ammonium, nitrate or in amide forms. Likewise, in the case of P fertilizers it may be water soluble, citrate-soluble or water and citrate-insoluble form. It is seen that N and K salts have much higher salt indices and hence, are more detrimental to germination than phosphatic salts when placed closer to or into contact with seed.

2. **Time of application:** The time of applying fertilizers has been found important only in the case of N-fertilizers which have a tendency to leach with irrigation or rains. In the case of P and K fertilizers, all the quantity applied at sowing has given the best results with most of the crops. In general, the crops which are grown in the rainy season should receive N-fertilizers in split doses so that the leaching of the nutrients in heavy rains may be avoided and an adequate supply of the nutrients at the critical stages may be assured.

3. **Method of application:** The placement of P-fertilizers has been found to be beneficial almost universally. The response to P fertilizers found much higher when the fertilizers placed in a band 5 cm wide to the side and 5 cm below the seed than that from broadcast application of the fertilizers. The introduction of high analysis fertilizers like urea, and the low volume sprayers and the use of aircraft have helped in experimentation involving the foliar application of fertilizers. For the effectiveness of the foliar application of nitrogen, it is essential that the crop should form a canopy, so that the nitrogen may be retained on the leaves.

4. **Use of nitrate inhibitors:** The development of nitrate inhibitors that could be added to the existing cheap nitrogen sources in minute quantities to control the nitrate release by decreasing the activity of nitrifying bacteria has been, therefore, one of the most significant developments in the field of increasing the efficiency of nitrogenous fertilizers. The commonest NO_3 inhibitors are:

   1. Oxamide (NH_2-COCO-NH_2O-31 % N).
   2. Dicyndiamide-NH_2C(=NH)NHCN-42 % N
   3. Thiourea – 36% N
   4. Urea-pyrolyzate-48 % N

These inhibitors seem to be promising as slowly available nitrogen releasing compounds. Two nitrification inhibitors, AM (2 amino-4 Chloro 6 Methyl Pyrimidin) and N-Serve (2 Chloro 6 trichloro methyl pyridine), blended with urea have been very effectively utilized in increasing the efficiency of N from urea.

1. **Use of chelating substance:** In the case of micronutrients, low solubility has been found to be the greatest limiting factor. The development of chelates of some of the important micro-organic complex which, although soluble themselves, do not ionize to any degree. They retain the metals in soluble form, permitting their absorption by the plants, yet preventing their conversion into insoluble form in the soil. The metallic ions commercially chelated are Fe, Cu, Zn and Mn. Numerous substances have the ability to chelate them. Four of the most important compounds found useful in agriculture are: 1 EDTA, 2. DTPA, 3. CDTA, 4. EDDHA

The exact mechanism of metal-chelate absorption and utilization by plants is not yet completely understood.

Tips for better fertilizer use efficiency

1. The fertiliser scheduling must be based on soil test.
2. Selection of fertilisers should be done according to the soil reaction *viz.*, acidic fertiliser for alkaline soils and basic fertilisers for acidic soil reactions.
3. Surface application through broadcasting should not be adopted but the fertiliser should be placed about 3-4 cms by the side or below the seed. This discourages weed growth also.

   The phosphatic and potassic fertilisers should be basal placed, because their poor mobility restricts them to the place of application. Therefore, they must be placed in the root zone.
4. Home mixing of fertilisers should be in accordance with the fertilizer-mixing guide and such fertiliser mixture must be applied as soon as possible.
5. In case of heavy soil type, half of the nitrogenous fertilisers should be basal placed and rest should be top-dressed in one split only.
6. But in case of light soils, nitrogen should be applied in three equal splits *i.e* 1/3 as basal, 1/3 after 30 days of sowing and the balance 1/3 about 50-60 days after sowing.
7. For at least a week, flooding with too deep water or poor drainage should be avoided after application of the fertilisers.
8. Top dressing should be done after draining out the water and weeding so that the loss of nutrient is minimum.Paddy fields, used for transplanting, should be puddled and fertilisers should be applied at the time of puddling. This will help fertilisers to penetrate and get stored in the soil.
9. The acidic soils should be treated with liming materials as and when required.
10. Deep placement of fertiliser, along with foliar feeding of nitrogen (*i.e.,* urea) through spraying of nitrogenous fertiliser in place of top dressing should be done in case of dry lands.
11. Addition of organic manures or green manuring should be done at least once in 3-5 years.Weed growth should not be permitted in cropped areas, during any part of the year.
12. In case of flooded fields or calcarious soils, use of slow release nitrogenous fertilisers like sulphur coated urea, urea super granules, neem coated or neem blended ureas should be used so that loss of nitrogen can be minimised.

13. Mud bolls, contain urea and should be used in case of deepwater crops because they help in proper placement and also reduce the loss of nitrogen from the field.
14. Appropriate plant protection measures and proper tillage practices should be adopted so that plants remain healthy and absorb the applied nutrients from the field.

FULL NAMES OF FERTILIZERS FACTORIES:

1. CLF : Coromandel Fertilizers Ltd.
2. DMCC : Dharamsi Morarji Chemical Co. Ltd.
3. FACT : Fertilizers and Chemicals Travancore Ltd.
4. FCI : Fertilizer Corporation of India Ltd.
5. GNFC : Gujarat Narmada Valley Fertilizer Co. Ltd.
6. GSFC : Gujarat State Fertilizers and Chemicals Ltd.
7. HCL : Hindustan Copper Ltd.
8. HFCL : Hindustan Fertilizer Corporation Ltd.
9. HZL : Hindustan Zinc Ltd.
10. GFC : Godavari Fertilizers and Chemicals
11. IEL : Indian Explosives Ltd.
12. IFFCO : Indian Farmer Fertilizers Co-operative Ltd.
13. IISCO : Indian Iron and Steel Co. Ltd.
14. JCF : Jayshree Chemicals and Fertilizers
15. KRIBHCO : Krishak Bharati Co-operative Ltd.
16. MCFL : Mangalore Chemicals and Fertilizers Ltd.
17. MFL : Madras Fertilizers Ltd.
18. MMTG : Minerals and Metals Trading Corporation
19. NFL : National Fertilizer Ltd.
20. PFL : Pradeep Fertilizers Ltd.
21. NLC : Neyveli Lignite Corporation.
22. PICUP : Pradeshiya Industrial and Investment Corporation of Uttar Pradesh
23. RCFL : Rashtriya Chemicals and Fertilizers Ltd.
24. SAIL : Steel Authority of India Ltd.
25. SFC : Shreeram Fertilizers and Chemicals

26. SPIC   : Southern Petrochemicals Industries Co-operative Ltd.
27. TISCO   : Tata Iron and steel Co. Ltd.
28. ZACL   : Zuary Agro Chemicals Ltd.

CHAPTER 9

METHODS OF FERTILIZER APPLICATION

The application of fertilizers to crop depends on type of fertilizer, nature of crops, age and stage of crops, crop response, soil types, soil reaction, methods of fertilizer application *etc*. hence, it is essential to know the different methods of fertilizers application to increase crop productivity by increasing fertilizer use efficiency.

9.1. TIME OF APPLICATION

The correct time of application is aimed at providing nutrients in sufficient quantities to meet the crop demand and at the same time avoiding excess availability and leaching losses. The time of application mainly depends on crop uptake pattern, soil properties, nature of the fertilizer material and utilization of carbohydrates.

(i) *Crop uptake*

Nitrogen, phosphorus and potassium are taken in large quantities in early stages of crop growth. For example: 93, 87 and 66 per cent of N, P, K uptake is completed by panicle initiation stage in finger millet. Nitrogen is necessary for the synthesis of proteins which are essential for the development of tissues. After flowering, most of the crops, especially cereals, contain lesser percentage of nitrogen due to greater accumulation of carbohydrates. The uptake of nitrogen is slow at the later stages, which is generally met from the soil by mineralization. Legumes require nitrogen until root nodules are formed. Potassium is taken gradually throughout the growth and development of the crop.

(ii) Soil properties and nature of fertilizers

Nitrogenous fertilizers are soluble and highly mobile in soil. Nitrogenous fertilizers are lost into deeper layers beyond root zone if the entire quantity of fertilizer is applied especially in light textured soils. Phosphatic fertilizers which are highly reactive are fixed in the soil and become immobile. Potassium fertilizers are less mobile since they are adsorbed on the clay complex. The entire quantity of phosphatic and potassium fertilizers are, therefore, applied in one dose at the time of sowing.

(iii) Utilization of carbohydrates

The level of carbohydrates and nitrogen in the plants are inversely related. When large quantities of nitrogenous fertilizers are applied, the level of carbohydrates in the plants decreases. With less nitrogen in plants, carbohydrate level in the plants increases. Under a sufficient level of nitrogen in the plant, carbohydrates are utilized for the synthesis of proteins. The assimilation of nitrogen requires energy which is obtained either from light or the breakdown of carbohydrates. The time of application of nitrogen, therefore, depends on the end product of the crop. In fodder crops, leafy succulent crop with higher level of proteins are preferred compared to fibrous crop with higher carbohydrates. Hence, application of nitrogen in several splits is necessary. If the fodder crop is grown for silage, it should have higher carbohydrates just before cutting for better quality silage. Application of nitrogen should be curtailed in the last stage. Similarly, nitrogen should not be applied for sugarcane during maturation as the economic product is carbohydrate.

(iv) Basal application

Application of fertilizers before or at the time of sowing is known as basal application. A portion of a recommended dose of nitrogen and entire quantity of phosphatic and potassium fertilizers are applied as basal. Modified forms of urea like urea super granules, sulphur-coated urea, neem-coated urea etc. are used for basal application.

(v) Split application

Application of recommended dose of fertilizers in two or three splits during crop period is known as split application of fertilizers. Application of fertilizers in thestanding crop is known as top dressing. The number of split applications has to be more in light soils and less in heavy soils. Nitrogen is applied in more splits for long duration crops. The stage of application is also important. In cereals, nitrogen is applied at tillering and panicle initiation stages in addition to basal application. Basal application is sufficient in pulses while for sugarcane, it is not necessary. Nitrogen is applied at 60, 90 and 120 days after sowing for sugarcane.

9.2. FACTORS INFLUENCING METHODS OF APPLICATION

(i) *Nature of the Soil*

Soil properties like texture, pH, CEC, nutrient and moisture status are important factors to be considered for selecting suitable method of application. Soil texture influences the mobility of the fertilizer material. Soil pH increases volatilization of ammoniacal fertilizers when it is more than 8 pH. The soils with high CEC retain the cations present in the fertilizer material and thus reduce leaching tosses. In soils low in phosphorus status, band placement of phosphatic fertilizer reduces fixation, while in soils of medium phosphorus status, incorporation of phosphorus fertilizer after broadcasting is better for higher availability. For crops grown on residual moisture, deep placement of fertilizer in the moist zone is essential and when it is not possible foliar application is resorted to.

(ii) *Nature of the Crop*

Depending on the type of root system and spacing adopted for the crop, different methods of fertilizer application are practiced. In crops with fibrous root system and those grown with closer spacing, most of top layer of the soil is occupied by the root system. In such a situation, broadcasting of fertilizer is resorted to followed by irrigation. In widely spaced crops with initial slow growth, point placement is adopted instead of broadcasting over the entire field.

(iii) *Nature of the fertilizer*

Suitable method of fertilizer application depends on the properties of fertilizers such as physical form, solubility and mobility. Mud-ball urea, pellets and briquetts of urea are amenable for placement with hand. Granules and prills can be drilled while granules, prills and powders can be broadcasted. Liquid fertilizers are applied with irrigation water alone or mixed with herbicide sprays. Soluble fertilizers can be applied as foliar application. Fertilizers containing plant nutrients which are immobile or less mobile are applied in the root zone. Fertilizers which are subject to volatilization and denitrification losses are incorporated into the soil.

9.3. METHODS OF FERTILIZER APPLICATION

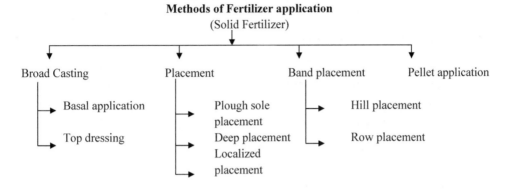

(A) Solid Fertilizers

1. **Broad casting:** Broadcasting is the method of application of fertilizer uniformly over the entire field by hand. It may be at the time planting or in standing crop as a top dressing. Broadcasting of fertilizers is of two types.

   i) **Broadcasting at sowing or planting (Basal application)**

   Basal application: The application of fertilizers at the time of sowing or planting is called basal application of fertilizers.

   Basal Dose: The dose (amount) of fertilizers applied at the time of sowing or planting is called basal dose.

   The main objectives of broadcasting the fertilizers at sowing time are to uniformly distribute the fertilizer over the entire field and to mix it with soil.

   Broadcasting at planting is adopted under certain conditions are as follows:

   a. Soils highly deficient, especially in nitrogen.

   b. Where fertilizers like basic slag, dicalcium phosphate, bone meal and rock phosphate are to be applied to acid soils, and

   c. When potassic fertilizers are to be applied to potash deficient soils.

   ii) **Top dressing:** The method of fertilizer application in the standing crops is known as top dressing. Usually, nitrate nitrogen fertilizers are top dressed in standing crops. Depending on the duration of the crop and soil type, top dressing may be more than one to meet the crop needs at times of greatest need of the crop. The objective of this method is to provide the nutrients, mainly nitrogen, in readily available form, for the growth of plants.

Topdressing with nitrogenous and potassic fertilizer should not be done when the leaves of plants are wet. This may burn or scorch the leaves if applied in the morning before 8 a.m. or just after rain.

Disadvantages of broadcasting

The main disadvantages of application of fertilizers through broadcasting are:

a. Nutrients cannot be fully utilized by plant roots as they move laterally over long distances.

b. The weed growth is stimulated all over the field.

c. Nutrients are fixed in the soil as they come in contact with a large mass of soil.

2. **Placement:** It is a method of placing fertilizer in the soil before sowing or after sowing the crops. Placement of phosphatic fertilizer below the seed, almost under all situations, has been found superior to broadcast application. Various methods of placement are:

   i) **Plough-sole placement:** The fertilizer is placed in the plough sole after opening the furrow with the plough, and this furrow is covered immediately as the next furrow is turned. This method has been recommended under following circumstances:

   - In dry soil where there is moisture only in the plough-sole layer.
   - In problematic soils where there is a problem of nutrient fixation more especially clay soils.
   - When the quantity of fertilizers to apply is small.
   - Development of the root system is poor.
   - Soils have a low level of fertility and to apply phosphatic and potassic fertilizer.

   ii) **Deep placement:** Deep placement is the method of fertilizer application, especially nitrogen, in the reduced zone to avoid nitrogen losses in low land rice. This method ensures better distribution of fertilizer in the root zone soil and prevents loss of nutrients by run-off and denitrification.

   iii) **Localized placement:** It is a method of placing fertilizers, into the soil, close to the seed or plant. The roots of young plants can get nutrients as per their requirement from the fertilizer applied by this method.

   Localized placement of fertilizers can be done by:

   a) **Contact placement or drill placement:** Drill placement refers to drilling seeds and fertilizer simultaneously at the time of sowing. In this method, the fertilizer is applied at the time of sowing by means of a seed-cum-fertilizer drill. This places fertilizer and the seed in the same row but at

different depths. This method has been found suitable for the application of phosphatic and potassic fertilizers in cereal crops, but sometimes germination of seeds and young plants may get damaged due to higher concentration of soluble salts if care is not taken. But this method is not suitable for pulse crops.

b) **Band placement:** In this method, the fertilizer is placed in bands on one side or both sides of the row, about 5 cm. away from the seed or plant in any direction (applying the fertilizer in continuous bands, close to the seed or plant). This method is ideal for crops grown in wide space *i.e.,* brinjal, papaya, sugarcane, tobacco, maize *etc.*

Band placement is of two types.

  i) **Hill placement:** It is practiced for the application of fertilizers in orchards. In this method, fertilizers are placed close to the plant in bands on one or both sides of the plant. The length and depth of the band varies with the nature of the crop.

  ii) **Row placement:** When the crops like sugarcane, potato, maize, cereals etc., are sown close together in rows, the fertilizer is applied in continuous bands on one or both sides of the row, which is known as row placement.

c) **Side dressing:** In this method, fertilizers are applied along the side of a row or around the plant and mixed into the soil with a spade (khurpi).

d) **Pellet placement:** It is application of fertilizer, especially nitrogen in pellet from in the low land rice avoiding nitrogen loss from applied fertilizer.

Small pellets of convenient size are made after mixing the nitrogenous fertilizers, specially, urea with soil (Usually 1 : 10 urea – soil ratio) and they are applied one to two inches deep between the rows of the paddy crop. The pellets are deposited in the soft mud of paddy fields. This method of fertilizer application decreases the nitrogen loss through leaching or by run-off of water.

Advantages of placement of fertilizers

The main advantages are as follows:

- When the fertilizer is placed, there is minimum contact between the soil and the fertilizer, and thus fixation of nutrients is greatly reduced.
- The weeds all over the field cannot make use of the fertilizers.
- Residual response of fertilizers is usually higher.
- Utilization of fertilizers by the plants is higher.
- Loss of nitrogen by leaching is reduced.
- Being immobile, phosphates are better utilized when placed.

SOIL FERTILITY AND NUTRIENT MANAGEMENT

(B) Liquid Fertilizers

Some fertilizers are also available in liquid form or prepared by dissolving water soluble solid fertilizers. Liquid form fertilizers are applied with irrigation water or for direct application. Quick response to the crop, high use efficiency, ease of handling, less labour requirement and possibility of mixing with herbicides has made the liquid fertilizers more acceptable to farmers.

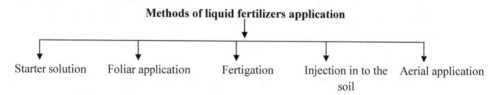

1. **Starter solution:** These are solutions of fertilizers prepared in low concentrations used for soaking seed, dipping roots or spraying on seedlings for early establishment and growth.

2. **Foliar application:** It refers to the spraying of fertilizer solutions containing one or more nutrients on the foliage of growing plants.

   Several nutrient elements are readily absorbed by leaves when they are dissolved in water and sprayed on them. The concentration of the spray solution has to be controlled; otherwise serious damage may result due to scorching of the leaves. Foliar application is effective for the application of minor nutrients like iron, copper, boron, zinc and manganese. Sometimes insecticides are also applied along with fertilizers.

   In the case of calcium, transport from roots to fruit is limited, so foliar applications are the best method we know of go get more calcium into fruit tissue to reduce post harvest disorders.

   If soil pH limits nutrient availability, and ground applied fertilizes are not taken up, foliar fertilizers may be a valid option. In this case, a soil sample should be taken to determine pH, and a leaf tissue sample taken to determine the need for addition foliar fertilization. In some cases poor root health from compaction, replant disease, crown rot, mouse damage, water logging or other problem may warrant foliar feeding of trees. However, the fertilizer in the required amount cannot be phototoxic as a foliar spray, and uptake must have been demonstrated with the product under consideration.

   Zinc uptake deserves special attention. In our soils zinc is largely immobile and it is difficult to supply roots with adequate amounts of available Zn. As a result of limited soil availability, zinc is applied as a foliar spray. Research has shown that only a small amount of Zn can be taken up by leaves, however foliar application are still more successful than soil applied Zn.

3. **Fertigation:** Fertigation refers to the application of water soluble fertilizers through irrigation water either in open or closed system of irrigation. The open system includes lined and unlined open ditches and gated pipes that are used for furrow and flood irrigation. Sprinkler and drip (trickle) systems are main closed systems. Nitrogen and sulphur are the principal nutrients applied by fertigation. In this system fertilizer solution is distributed evenly in irrigation. The availability of nutrients is very high therefore the efficiency is more. In this method liquid fertilizer as well as water soluble fertilizers are used. By this method, fertilizer use efficiency is increased from 80 to 90 per cent.

The fertigation allows applying the nutrients exactly and uniformly only to the wetted root volume, where the active roots are concentrated. This remarkably increases the efficiency in the application of the fertilizer, which allows reducing the amount of applied fertilizer. This not only reduces the production costs but also lessens the potential of groundwater pollution caused by the fertilizer leaching.

Advantages of the fertigation are:
- Fertigation saves the fertilizer, time, energy and labour.
- Higher absorption of nutrients by the plant. Because of nutrients and water are supplied near the active root zone through fertigation which results in greater absorption by the crops.
- Due to fertigation increase crop yield up to 25 to 50 per cent.
- Increase the fertilizer use efficiency. Fertilizer use efficiency through fertigation ranges between 80-90 per cent, which helps to save a minimum of 25 per cent of nutrients.
- By this method, convenient use of compound and ready-mix nutrient solutions containing also small concentrations of micronutrients which are otherwise very difficult to apply accurately to the soil.
- Minimize the water pollution. Because leaching losses of mobile nutrients are minimized through fertigation.

4. **Soil application (Inject in to the soil):** Liquid fertilizer such as anhydrous ammonia is applied directly to the soil with special injecting equipment. Liquid manures such as urine, sewage water and shed washing are directly let into the field.

5. **Aerial application:** In areas where ground application is not practicable, the fertilizer solutions are applied by aircraft particularly in hilly areas, in forest lands, in grass lands or in sugarcane fields etc.

Give the scientific reasons of the followings:

1. ***Phosphorus fertilizer use efficiency is very less:*** Because phosphorus **fixation** predominates in both acidic and alkaline soils, resulting in its low efficiency. In acid soils, it reacts with Fe and Al compounds and forms Iron and iluminium phosphate which are not available to plant. Whereas, in alkaline soils, it fix with Ca and Mg and forms insoluble compounds of calcium and magnesium phosphate which are also not available to plant. (***Fixation:*** *Readily available form of nutrients are converted in to unavailable form is called nutrient fixation*)

2. ***Fertigation increase the crop yield:***

   Due to following reasons fertigation increase crop yield:
   - It increases water and fertilizer use efficiency.
   - It minimize the leaching losses of nutrients with pressurized irrigation system (drip and sprinkler)
   - By fertigation, we can synchronize the nutrient needs to growing plant
   - Minimum fixation of nutrients with this method.
   - Minimize soil and water pollution.

3. **Split/foliar application of nitrogenous fertilizers should be recommended.**

   Firstly, nitrogen (especially nitrate form) is highly mobile in the soils and secondly, it is requires throughout the growing period of plant. Hence, to synchronize the crop need and to prevent leaching losses of nitrogen, the split/foliar application of nitrogenous fertilizers are advisable.

4. **Phosphatic fertilizers should be applied as a basal dose.**

   Firstly, phosphorus compounds are immobile in the soils and secondly, it requires in high amount at early state of plant growth for dell division and root development. Therefore, it is advised to apply phosphatic fertilizers are as a basal dose.

Nutrient interactions

Interactions may be defined as an influence, a mutual or reciprocal action of one element on another in relation to plant growth. The interaction effect may be positive or negative or it may have no any effect. The problem of zinc phosphate interaction is too well-known. An experiment conducted at Ludhiana, it was observed that as the amount of phosphate increased beyond a certain level-50 kg P_2O_5 in this case, it depressed the yield and reduced the beneficial effect of zinc.

CHAPTER 10

PROBLEMATIC SOILS

The soils which are unfavourable for cultivation of field crops because of one or more unfavourable soil properties/characteristics (viz. Soluble salts, soil reaction, ESP, water logging, aeration *etc.*) are adversely affect the optimum soil productivity is called problematic soils. The problematic soils need to be classified in to various groups for developing special systems of management for specific types of problems and constraints in the production of crops. The major problematic soils of India are as follows.

Table 10.1: Major problematic soils of India

Sr. No.	Problematic Soils	Key Diagnosis	Major constraints
1.	Clay soils	Dominated by clay particles	Water logging, compaction, poor aeration, difficult to cultivate
2.	Sandy soils	Dominated by Coarse sand particles	poor fertility, low SOM, low water holding capacity, erosion
3.	Acid soils	Soil pH is less than 6.5	Fe, Al toxicity (Strong acid soil)
4.	Salt affected soils		
	a. Saline soils	ECe is greater than 4.0 dS/m	High osmotic potential, nutrient imbalance
	b. Sodic soils	ESP is greater than 15	Deteriorated physical condition, Na toxicity, nutrient imbalance
	c. Saline sodic soils	ECe is greater than 4.0 dS/m and ESP is greater than 15	High osmotic potential, deteriorated physical condition, nutrient imbalance

[Table Contd.

Contd. Table]

Sr. No.	Problematic Soils	Key Diagnosis	Major constraints
7.	Calcareous soils	$CaCO_3$ is greater than 5.0 %	P, Fe deficiency
8.	Water logged soils	Water Stagnation, Low infiltration rate,	Poor aeration
9.	Degraded soils	Based on soil analysis	–
10.	Compacted soils	High bulk density	Poor aeration, poor root penetration, water logging
11.	Impermeable soils	Low hydraulic conductivity (HC) and infiltration rate	Poor aeration, water logging

Obtaining the maximum production potential of a particular crop depends on the growing season environment and the skill of the producer to identify and eliminate or minimize factors that reduce yield potential. More than 50 factors affect crop growth and yield potential. (*i.e.* climatic, Adaphic and genetic). Although, the producer cannot control many of the climate factors, most of the soil and crop factors can and must be managed to maximize productivity. Therefore, more emphasis is given to soil related constraints because majority of soil related problems can be altered in to favourable condition to maximize the soil productivity.

10.1 SALT AFFECTED SOILS

Soils, in which concentration of *salts* is so high as to adversely affect plant growth and crop productivity, are called salt affected soils. Some amounts of salts are always present in the soil. When the concentration of these salts is low, they are not harmful for the growth of plants. But with the increase in salt content of the soil to high levels, the plant growth adversely affected which, in turn, decreases the productivity of agricultural crops. The extent of reduction in growth and decrement in productivity, however, depend upon many factors such as kind and content of salt constituents, soil texture, distribution of salts in the soil profile, the species of plant grown, level of soil - water - crop management and climatic condition.

Classification of salt affected soils

The salt affected soils needs to classified in to various groups for developing special systems of management for specific types of problems and constraints in the production of crops. In 1954, the US Salinity Laboratory Staff grouped salt affected soils in to three distinct classes based on the behavior of salts in the soils viz. (i) Saline soils, (ii) Alkali soils and (iii) Saline- alkali soils.

A. Saline Soils (Synonymous: Solonchak (Russian term), Saline non sodic, White alkali)

Saline soils contains sufficient concentration of soluble salts in the root zone soil which are adversely affects the crop productivity OR simply, the accumulation of water soluble salts in the soil which restrict the crop production is called saline soil. The amount of soluble salts present in the soil is determined by the electrical conductivity or individual analysis of salts present in the soil. Among the salts present in the soil, Ca, Mg, Na and K are the dominant cations whereas CO_2, CO_3, Cl, SO_4 are the dominant anions in arid and semi arid region of the world. The process of accumulation of soluble salts in the soils is known as *salinization*.

Soluble salts: Salts which dissolved in soil water and are free to move with the soil water

Soil solution: The liquid phase of soil, consisting of soil water also contains dissolved salts and thus it is called soil solution

Causes of Salinization

Salinization or the accumulation of the salts occurs in the following ways:

1. **Primary minerals:** It is the original and important direct source of all the salt constituents. During the process of weathering, which involves hydrolysis, hydration, solution, oxidation and carbonation various constituents like Ca, Mg and Na are gradually released and made soluble. Eg. Halite (NaCl), Calcite ($CaCO_3$), Dolomite [Ca Mg $(CO_3)_2$].

2. **Arid and semi-arid climate:** Salt affected soils are mostly formed in arid and semi-arid climate where low rainfall and high evaporation prevails. The low rainfall in these regions is not sufficient to leach out the soluble weathered products and hence the salt accumulates in the soils. Further, high evaporation in these areas, lead to accumulate salt in the root zone due to capillary rise of salt with evaporating water from the lower zone. The intensity of salinization increases with increases in dryness of the climate. Salt affected soils in humid region exists only in areas subjected to sea water intrusions in deltanic regions and other low lying areas along the sea cost which get inundated by the sea water.

3. **Sea as a source of salts :** The ocean may be the source of salts as in soils where the parent materials consists of marine deposits that were laid down during earlier geological periods and have since been uplifted. The ocean is

also the source of the salts in low-lying area along the margin of seacoasts. Sometimes salts is moved inland through the transportation of spray by winds be called "cyclic salts". In Gujarat salts affected soils observed in *Bhal* region is due to marine deposit of receding of the sea while seawater inundation is the cause in several coastal regions in south Gujarat.

4. **Restricted leaching and transportation:** In arid regions, the leaching and transportation of soluble salts to ocean is not as complete as in humid regions. In arid regions the leaching is usually localized in nature and soluble salts may not be transported far. This occur not only because there is less rainfall available to leach and transport the salt but also because of the high evaporation rates which tend to concentrate the salt in the soils and in surface water.

5. **Low permeability of the soil:** This causes poor drainage by impeding the downward movement of water. Low permeability may be results of an unfavorable soil texture or structure or hard pan/clay pan, because of the low permeability the ground water table may raise or because of continuous deposition of soluble salts in the soils.

6. **Ground water:** Ground water contains large amounts of water soluble salts which depend upon the nature and properties of the geological material with which water remains in contact where water table and evapotranspiration rate is high, salts along with water move upward through capillary activity and salts accumulation on the soil surface.

7. **Irrigation water:** The application of irrigation water without proper management (i.e. lack of drainage and leaching facilities) increases the water table and surface salt content in the soils.

8. **Poor drainage of soil:** During the periods of high rainfall, the salts are leached from the upper layer and if the drainage is impeded, they accumulate in the lower layer. When the water evaporates the salts are left in the soils. Such soils are generally developed in low-lying areas.

9. **High water table:** The ground waters of arid regions usually contain considerable quantities of soluble salts. If the water table is high, large amount of water moves to the surface by capillary action and evaporated, leaving soluble salts on the surface.

10. **Canal as a source of salinization:** Although, canal water practically contain very little amount of soluble salts, but during earlier stages, the excess use of canal water hastens the rise of ground water table. When, the water table rises within 5 or 6 feet from the soil surface, the ground water move upward into the root zone and to the soil surface. Under such condition, ground water as well as irrigation water contributes to salinization of soils.

B. Alkali Soils (Synonymous: Solonetz (Russian term), Non saline sodic, sodic, Black alkali)

Alkali soils have sufficient sodium saturation on the exchange complex and alkalinity to adversely affect plant growth and crop productivity. Carbonates (CO_3^- + HCO_3^-) of sodium are dominant salts. The concentration of natural salts (Cl^- and SO_4^-) is much lower.

Alkalinity or Alkaline: It indicate the reaction of soil, means soils contains excess alkalinity (pH more than 7.00).

Alkali: It indicates condition of soil where alkali ion (sodium) is dominant on exchange complex of the soil.

Alkalization: It is the process of accumulation of sodium ion on soil exchange complex is known as alkalization.

Causes of Alkalinity:

Process where exchangeable Na content in soil increased due to precipitation of Ca and Mg as carbonate (Na_2CO_3 or $NaHCO_3$) by low of mass action, Ca and Mg replaced by Na on exchange complex.

1. **Hydrolysis of sodium silicate or weathering of minerals.**
   Na X + H_2O — hydrolysis → NaOH + H X
   NaOH + CO_2 — hydrolysis → $NaHCO_3$ →
   $2NaHCO_3$ — decomposition → Na_2CO_3 + CO_2 + H_2O
   Na_2SiO_3 → Sodium silicate → highly sodic
   NaOH → also highly sodic

2. **Replace Na_2SO_4 or NaCl by $CaCO_3$**
   Na_2SO_4 + $CaCO_3$ → Na_2CO_3 + $CaSO_4$
   $2NaCl$ + $CaCO_3$ → Na_2CO_3 + $CaCl_2$

3. **Hydrolysis of exchangeable Na**
   Na X Na + H_2O + CO_2 → H X H + Na_2CO_3
   Na X Na + $CaCO_3$ → Ca X Ca + Na_2CO_3
   Na X Na + $Ca(HCO_3)_2$ → Ca X Ca + $NaHCO_3$

4. Reduction of Na_2SO_4 through microorganisms
   Na_2SO_4 + 2C + CO_2 → Na_2S + $2CO_2$
   Na_2S + CO_2 + H_2O → H_2S + Na_2CO_3

5. **Decomposition of organic matter**

$Na_2SO_4 \rightarrow Na_2S + 2O_2$

$Na_2S + 2H_2O \rightarrow H_2S + 2NaOH$

$NaOH + CO_2 \rightarrow NaHCO_3$

6. **Use of alkali or sodic water for irrigation**
7. **Excessive use of basic fertilizers:** Use of basic fertilizers like Na_2NO_3, basic slag, etc. may develop alkalinity in the soils.
8. **Humid and semi-humid regions :** Alkaline soils develop in other area also e.g. in semi-humid and temperate regions especially in depressions where drainage is defective and where the underground water table is high or close to the surface.

Distribution of Salt Affected Soils

World - 952 million hectares

India - 7.02 million hectares

Gujarat - 1.12 million hectares

In India the state wise estimated area having salt affected soils (C.S.S.R.I., Karnal) is as follows It is seen from the above data that Gujarat has second largest salinized area.

Table 10.1: State wise distribution of salt affected soils

States	Area (Million ha)	States	Area (Million ha)
Uttar Pradesh	1.295	Orissa	0.404
Gujarat	1.214	Maharashtra	0.534
West Bengal	0.850	Karnataka	0.404
Rajasthan	0.728	Madhya Pradesh	0.224
Punjab	0.688	Andhra Pradesh	0.042
Haryana	0.526	Other states	0.040

Morphological feature and Characterization of Salt Affected Soils

The characteristics of different salt affected soils are as under :

1. **Saline non-alkali soils :** These soils contain primarily the soluble salts of chlorides and sulphates of sodium, calcium and magnesium. The electrical

conductivity of the saturation extract (ECe) is greater than 4 dSm^{-1} and the exchangeable sodium percentage (ESP) is less than 15. These soils may contain small quantity of bicarbonates, but soluble carbonates are generally absent. Ordinarily, the pH is less than 8.5. The saline soils are often recognized by the presence of white crusts of salts on the surface. These soils correspond to Hilgard's "White alkali soils" and to the "Solonchaks" of the Russian soil scientists. Sometimes the solonchaks may be active or residual. The active solonchaks are connected to ground water through capillary fringe and the residual solonchaks or dry solonchaks are not at all connected to ground water and are observed in extreme arid locations.

- Accumulation of neutral salts
- Predominantly NaCl and Na_2SO_4. Na_2CO_3 is always absent in saline soils.
- In arid and semi-arid region
- High temperature and low rainfall
- Effect on plant : osmotic pressure increased, hence water come out from the plant into soil (reverse water uptake) and toxic effect of chlorides, borates and magnesium on plant
- Known as white alkali soils, salts accumulates on soil surface and look whitish.
- White salts on bunds and ridges of beds and irrigation channels

2. **Alkali soils :** The non-saline alkali soil is applied to soils for which the ESP is > 15 and ECe is < 4 dSm^{-1}. The pH usually ranges between 8.5 and 10. These soils correspond to HIlgard's "Black alkali" soils and in some cases to solonchaks as the latter term is used by the Russian scientists. They frequently occur in semi-arid and arid regions. The drainage and leaching of saline-alkali soils, in absence of gypsum, leads to the formation of such soils. Dispersed and dissolved organic matter present in the such soils may be deposited on the soil surface by evaporation, thus causing darkening and giving rise to the term "Black alkali" the characteristic morphological features of these soils are clay is highly dispersed and is may be transported downward through the soil and accumulated at lower depths. A few cm of the surface soil may be relatively coarse in texture and friable but below, where the clay accumulates, the soil may develop a dense layer of low permeability that may have the presence of carbonate ions, Ca and Mg are precipitated hence, the soil solutions of non-saline alkali soils usually contain only small amounts of these cations, sodium being the predominant one. ESP becomes high. Non-saline alkali soils in some areas of western United States have ESP considerably above 15, and yet the pH readings, especially in the surface soil, may be as

low as 6. These soils have been referred to by De sigmond as "Degraded alkali soils". The physical properties, however, are dominated by the exchangeable Na and are typically those of non-saline alkali soils.

- Known as black alkali due to dissolved organic matter
- Dominant salts are Na_2CO_3 and Na_2HCO_3
- Destroy physical properties of soil

3. **Saline- alkali soils :** These soils are formed as a result of salinization and alkalinization and have high amount of soluble salts and appreciable quantity of sodium on the exchange complex. The ESP of these soils is greater than 15 and the ECe is more than 4 dSm^{-1}. As long as excess salts are present, the appearance and properties of these soils are generally similar to those of saline soils. The pH readings are seldom higher than 8.5 and particles remain in flocculate conditions. If the excess salts are leached down, the properties of these soils may change markedly and became similar to those of non-saline alkali soils. As the concentration of the salts in the soil solution is lowered, some of the exchangeable sodium hydrolyzes and forms NaOH. This may change in Na_2CO_3 upon carbonation reaction. Soil became strongly alkali (pH > 8.5), the particles disperse and the soil becomes unfavourable for the entry and movement of water and for tillage. The management of saline sodic soils continues to be a problem until the excess salts and exchangeable sodium are removed from the root zone and a favourable physical condition of the soil is reestablished.

- Both the type (Saline and Sodic) of characteristic are observed

4. **Degraded alkali or Sodic Soil :** If the extensive leaching of a saline-sodic soil occurs in the absence of any source of calcium or magnesium, part of the exchangeable sodium is gradually replaced by hydrogen. The resulting soil may be slightly acid with unstable structure. Such a soil is called degraded alkali or sodic soil.

$$\text{clay] Na} + H_2O \leftrightarrow H \text{ [clay} + NaOH \downarrow$$

(Acid soil on the leaching surface horizon)

$$2NaOH + CO_2 = Na_2CO_3 + H_2O$$

(from soil)   (Alkali soil in the sub-surface horizon)

Sodium carbonate (Na_2CO_3) dissolves humus and is deposited in the lower layer. The lower layer thus acquires a black colour. At the same time H-clay formed in this way does not remain stable. The process of break down of H-clay under alkaline condition is known as solodisation.

Chemical characteristics

1. The soil pH of surface soil is acidic (pH 6.0). This layer is usually very thin, hardly a few inches in depth.
2. The lower layer which constitutes the main body has a high pH (more than 8.5).
3. ESP is greater than 15 and ECe less than 4 dS/m.
4. The lower layer has black colour and develops prism like structure.
5. Soils became compact and have low infiltration and permeability.

Classification of salt-affected soils used by the Natural Resources Conservation Service (NRCS)

soil	ECe (dSm)	pHs	ESP	SAR	SoilStructure
Non Saline	< 4.0	6.5-7.0	<15	<13	flocculated
Saline	> 4.0	<8.2	<15	<13	flocculated
Sodic	< 4.0	8.2 -10.0	>15	>13	disperse
Saline -Sodic	> 4.0	8.2-10.0	>15	>13	flocculated

Characterization of salt affected soils

Characteristics	Saline soil	Alkali soils	Saline-alkali soils	Degraded alkali soil
Content in soil	Excess soluble salts	Presence of exchangeable sodium on the soil complex.	Soil contains Na-clay as well as soluble salts.	Hydrogen (H^+) ions in the upper layer and sodium (Na^+) in the lower layer.
ECe (dS/m)	> 4	< 4	> 4	–
Soil pH	Less than 8.5	8.5-10	More than 8.5	pH about 6 in the surface soil and pH 8.5 in the lower layer
ESP	Less than 15	More than 15	More than 15	More than 15
Sodium adsorption ratio (SAR)	Less than 13	More than 13	More than 13	Less than 13 in the surface and greater than 13 in the lower horizon
Total soluble salt content	More than 0.1 %.	Less than 0.1 %	More than 0.1 %	Less than 0.1 %

[Table Contd.

Contd. Table]

Characteristics	Saline soil	Alkali soils	Saline-alkali soils	Degraded alkali soil
Dominant salts	Sulphate (SO_4^{2-}), chloride (Cl^-) and nitrates (NO_3^-)	Sodium carbonate (Na_2CO_3)	–	Sodium carbonate (Na_2CO_3) in lower layer
Organic matter content	Slightly less than normal soils	Very low due to the presence of sodium carbonate (Na_2CO_3) barren (Usar)	Variable	Low
Colour	White	Black	–	Black in lower layer
Physical condition of the soil	Flocculated condition, permeable to water and air. Soil structure optimum.	Deflocculated condition, permeability to water and air is poor. Very poor soil structure.	Flocculated or deflocculated depending upon the presence of sodium salts and Na-clay	Compact low infiltration and permeability. It develops columnar structure
Other name	White alkali (Solonchak)	Black alkali (Solonetz)	Usar	Solod, Soloth

Reason: *Saline soil is also known as white alkali soils* because these soils are characterized by saline efflorescence or white encrustation of salt at the surface. (Simply, Saline soil has a surface crust of white salts due to capillary rise of salts with evaporating water during summer.)

Alkali soil is also known as black alkali soil. Due to presence of high amount of exchangeable sodium and high pH, the soil colloids get dispersed and organic matter present in the soil are dispersed and dissolved. When these dispersed and dissolved organic matter is deposited in the surface, alkali soils give dark brown - black appearance.

Problems of salt affected soils

1. **Saline Soils:** There are various problems of saline soils that interfere with the plant growth.

(i) Soils are usually barren but potentially productive. Wilting coefficient of saline soil is very high. and amount of available soil moisture is low.

(ii) Absorption of water and nutrients : Excessive salts in the soil solution increase the osmotic pressure of soil solution in comparison to cell sap. This osmotic effect increases the potential forces that hold water in the soil and makes it more difficult for plant roots to extract moisture. During a drying period, salt in soil solutions may be so concentrated as to kill plants by pulling water from them (ex-osmosis). Due to high salt concentration plants have to spent more energy to absorb water and to exclude salt from metabolically active sites. At the same time various nutrient elements become unavailable to plants.

(iii) Salt toxicity : When the concentration of soluble salts increases to a high level then it produces toxic effect directly to plants such as root injury, inhibition of seed germination, etc.

2. **Alkali or Sodic Soils:** Excess exchangeable sodium in alkali soils affects both the physical and chemical properties of soils.

   (i) Dispersion of soil colloids : Under alkali conditions, the damage is not due to salt concentration. The influence of exchangeable sodium on the over-all physical properties of soils is associated mainly with the behaviour of the clay and organic matter, in which most of the cation exchange capacity (CEC) is concentrated. The sodium (Na^+) ion adsorbed by clay colloids causes deflocculation or dispersion of clay, which result in a loss of desirable soil structure, and helps for the development of compact soil.

   (ii) Other physical properties : Due to dispersion and compactness of soil, aeration, hydraulic conductivity, drainage and microbial activity are reduced.

   (iii) Caustic influence : It results high sodicity caused by the sodium carbonate (Na_2CO_3) and bicarbonate ($NaHCO_3$).

   (iv) Concentration of hydroxyl (OH^-) ion : High hydroxyl (OH^-) ion concentration no doubt has direct detrimental effect on plants. Damage from hydroxyl ions occurs at pH 10.5 or higher.

   (v) Specific ion effect : The presence of excess sodium in sodic soils may induce deficiencies of other cation like calcium (Ca^{2+}) and magnesium (Mg^{2+}). The action of sodium in inducing deficiencies of Ca^{2+} and Mg^{2+} appears to be three fold : (i) because sodium is comparatively loosely held in exchangeable form, the ions released to the soil solution in a fractional exchange are mostly sodium ions if the soil has a high exchangeable sodium percentage, (ii) at the high pH values (sodic soil) usually associated

with excess exchangeable sodium in the absence of excess salts, the soil solution contains bicarbonate and carbonate ions that tend to form insoluble precipitates of calcium and magnesium carbonates as follows :

$$Ca^{2+} + CO_3^{2+} = CaCO_3$$
(Soil solution) (Soil solution)   (Insoluble precipitation)

and (iii) exclusion of calcium and magnesium from absorption on competitive basis.

(vi) Availability of plant nutrients : The high pH in alkali or sodic soils decreases the availability of many plant nutrients like P, N, Mg, Fe, Cu, and Zn.

Appraisal/evaluation of Saline and Sodic Soils

A. Saline Soils

Different criteria are employed for characterizing soil salinity and those are given below:

1. **Soluble salt concentration in the soil solution:** In saline soils, the water soluble salts concentration in the soil solution is very high and as a result the osmotic pressure of the soil solution is also very high. As a result of which the plant growth is affected due to wilting and nutrient deficiency. Salt content more than 0.1 % is injurious to plant growth.

2. **Osmotic Pressure (O.P.):** It should be assessed at field capacity soil moisture regimes. Besides the relation between OP and electrical conductivity (EC) for salt mixtures found in saline soils, is given below:

   OP (in atmospheres or bars) = 0.36 X EC Where; EC expressed as dS/m

3. **Electrical conductivity (EC) of the soil saturation extract:** Measurement of EC of the soil saturation extract (ECe) is also essential for the assessment of the saline soil for the plant growth and is expressed as dS/m (formerly mmhos/cm).

4. **Concentration of water-soluble boron:** The determination of water-soluble boron concentration is also another criterion for characterization of saline soils. The critical limit of boron concentration for the plant growth is given below:

Class		Boron concentration (ppm)
I	Safe	< 0.7
II	Marginal	0.7 – 1.5
III	Unsafe	> 1.5

SOIL FERTILITY AND NUTRIENT MANAGEMENT

B. Alkali soils

There are various methods employed for its approval that are as follows:

(i) Exchangeable sodium percentage :

$$\text{ESP} = \frac{\text{Exchangeable Na (me/100 g soil)}}{\text{CEC (me/100 g soil)}} \times 100$$

Sometimes soil pH also gives an indication of soil alkalinity indirectly. It is generally found that the higher the ESP, the higher is the soil pH.

(ii) Sodium adsorption ratio : The U.S. Salinity Laboratory developed the concept of Sodium Adsorption Ratio (SAR) to define the equilibrium between soluble and exchangeable cations as follows :

$$\text{SAR} = \frac{[Na^+]}{\sqrt{\frac{Ca^{+2} + Mg^{+2}}{2}}}$$

(Where, the concentration of Na^+, Ca^{2+}, and Mg^{2+} of Saturation extract are expressed in me/l)

The value of SAR can be also used for the determination of Exchangeable Sodium Percentage (ESP) of the saturation extract by using the following formula:

$$\text{ESP} : \frac{100\,(-0.0126 + 0.01475\ \text{SAR})}{1 + (-0.0126 + 0.01475\ \text{SAR})}$$

Sometimes the following regression equation is used for the appraisal of alkali soil by determining the value of ESP from the value of SAR.

Y = 0.0673 + 0.035 X Where; Y indicates ESP and X indicates SAR

Soils having SAR value greater than 13 are considered as alkali or sodic

Causes of poor growth on saline soils

The crop growth on salt affected soils is poor due to one or another reason. The various reasons given for poor crop growth under such conditions are discussed below:

(A) Water availability theory: Due to high salt concentration plants have to spent more energy to absorb water and to exclude salt from metabolically active sites. At the same time various nutrient elements become unavailable to plants.

(B) Osmotic inhibition theory: According to osmotic inhibition theory, plant growth is inhibited by the excess of solute taken up from saline media. The osmotic inhibition theory thus postulates that the salts act inside the plants, but it does not specify how the inhibition of growth is effected. The inhibition could even result in even part from water deficiency in a sense different from that envisioned by the water availability theory. The presence of excess solutes in the plant decreases the free energy of unit mass of water even though the absolute mass of water in the excess of salts present externally.

(C) Specific toxicity theory: According to the specific toxicity theory, soil salinity exerts a detrimental effect on plants through the toxicity of one or more specific ions (cations as well as anions) in the salts present in excess. Accordingly, there may be toxicity of chlorides, bicarbonates and boron and to a lesser concentration of magnesium or its salts may also induce calcium deficiency.

Causes of poor crop growth on alkali soils

The reasons of low crop production on such soils are as follows:

(A) Adverse physical conditions: The alkali soils have poor physical conditions. The permeability of air and water and the hydraulic conductivity are at a lower most state due to breakdown of aggregates and dispersion of individual clay-colloids. The breakdown of aggregates is due to dissolution of organic matter, which acts as a cementing agent for binding individual clay particles, due to formation of alkali solution. The dispersed clay plugs all the macro and micro capillaries thereby hampering the movement of air and water. Such dispersed clay swells considerably due to high hydration capacity of Na ions and remain unflocculated in presence of water. The downward movement of water is practically zero and hence they remain waterlogged when irrigation is given or water is added through rain. On drying, such soils form very large clods, which are very hard in nature. The hard crust formation on the surface of the soil is most common characteristics. The tillage operations are very difficult to carry due to increase in bulk density, which is due to deflocculation of clay. Because of very adverse physical conditions, the germination as well as the root growth is considerably reduced. Because of this reduction, the overall crop growth is not at all satisfactory.

(B) High sodium on exchange complex: Excess amount of sodium reduces the crop growth considerably *i.e.,* there arises sodium toxicity because of excess concentration of Na. The plants exhibit the deficiency of Ca and to some extent Mg due to well-known principle complementary ion effect i.e. higher concentration of Na would reduce the uptake of Ca^{++} and Mg^{++}. The relative concentrations of Ca and Mg in such soils are low due to precipitation of these ions during process of alkalization. The relationship such as Ca/Na, Mg/Na, (Ca + Mg)/Na or K/Na are well-known and antagonistic relationships.

(C) **Effect of high pH:** The solution soils have high pH and the values range from 8.5 to as high as 10 or 11. The high pH reduces the availability of P, Zn, Cu, Mn and Fe (the availability of P increases at a very high pH value, 12 and above, due to formation of soluble sodium phosphate). The microbial activity is also at standstill due to unfavourable pH and the processes of mineralization, ammonification or nitrification are practically negligible. Apart from nutritional deficiency due to high pH, the higher concentration of OH ion itself is not favorable to crop growth.

In case of saline-alkali soils excess sodium may reduce the crop growth. However, the physical conditions of such soils are not disturbed due to excess amount of salts.

Methods of reclamation

I. **Mechanical**
- Construction of embankment to prevent tidal see water
- Land leveling and contour bunding
- Establishment of drainage network
- Breaking of hardpan in the subsurface layer through boring auger hole

II. **Hydrological**
- Flushing
- Leaching
- Drainage

III. **Chemical**
- Use of amendments

IV. **Physical**
- Scrapping of salt crust
- Deep tillage, sub soiling, profile inversion
- Use of soil conditioners e.g. sand, *tanch*, ash, manures and synthetic polymers like PVAC, PAM, and PVPC

V. Biological

- Agroforestry system
- Use of manures
- Green manure
- Selection of salt tolerant crops after afforestation

Hydrological method

1. **Flushing:** The salts can be removed by flushing which is the surface washing out of salts with the runoff water, which is collected at the sloppy end of the field. One serious drawback with flushing, as a means of removing soluble salts is its inability to flush through the soil and the salts have to come up to the surface for being removed through flushes. The flushing method is employed where moisture transmission characteristics into the profile are extremely poor. The USSSL does not advocate flushing as an effective means for washing out the salts.

2. **Leaching:** Leaching is the process of dissolving and transporting soluble salts by the downward movement of water through the soil. The leaching may be done by two methods viz., (i) flooding and (ii) sprinkler methods. The flooding method requires more amount of water and removes the salt at a greater depth. The sprinkler method has been found efficient in removing the salts from top layer (60 cm) with less amount of water. The method to be adopted for reclamation depends on (i) the crop to be grown, (ii) topography, (iii) soil characteristics, (iv) availability of water, (v) depth of under ground water table and (vi) magnitude of salinity/sodicity in the soil.

3. **Drainage:** Drainage in agriculture is the process of removal of excess water from soil. Excess water discharged by flow over the soil surface is referred to as surface drainage, and flow through the soil is termed internal or subsurface drainage. The term "artificial drainage" and "natural drainage" indicate whether or not man has changed or influenced the drainage process. The design of drainage systems is influenced by many factors which taken into consideration e.g. drainage requirement, water transmission properties of soil and boundary conditions, water application efficiency, physiography of land, etc. The types of relief drains are pumped wells, tile or open drains may serve of these purposes. The drain should be placed below 2 m depth with a distance of 25 to 75 m according to soil texture orienting perpendicular to the direction of ground water flow.

Chemical procedure – use of amendments

In case of saline sodic and sodic soils, the exchange complex is saturated to varying degree with Na. The reclamation procedure in such cases also involves the use of amendments for replacing exchangeable Na.

A. Different types of amendments

The chemical amendments used are as under :
1. Soluble calcium salts e.g.
   - (i) Calcium chloride ($CaCl_2.2H_2O$)
   - (ii) Gypsum ($CaSO_4.2H_2O$)
   - (iii) Calcium sulphate ($CaSO_4$)
2. Acid or acid formers e.g.
   - (i) Sulphur (S)
   - (ii) Sulphuric acid (H_2SO_4)
   - (iii) Iron sulphate ($FeSO_4.7H_2O$)
   - (iv) Aluminium sulphate ($Al_2(SO_4)_3.18H_2O$)
   - (v) Lime sulphur (calcium poly sulphide) (CaS_5)
   - (vi) Pyrites (FeS_2)
3. Calcium salt of low solubility
   - (i) Ground lime stone ($CaCO_3$)
   - (ii) By-product lime from sugar factories e.g. pressmud

The kind and amount of chemical amendment to be used for the replacement of exchangeable Na in soils depend upon the soil characteristics, the desired rate of replacement and economic considerations. Soluble calcium salts are preferred when soil does not contain alkaline earth carbonates or calcium carbonate. Acid or acid formers are preferred when soil contains alkaline earth carbonates or $CaCO_3$. Acid or acid formers are also used along with calcium salt of low solubility but the rate of reaction is very low.

B. Advantages and disadvantages of amendments

The $CaCl_2$ is highly soluble and Ca is readily available but its cost is a prohibitive factor. Iron and aluminium sulphates also hydrolyse readily in the soil to form H_2SO_4 but here also the cost is acting. Amendment, which can be used in calcareous soils but it requires special equipments and is hazardous in handling.

Sulphur is a slow acting amendment and large applications are needed. It requires more time for complete oxidation. In cool winter season, the oxidation rate is too slow to give satisfactory results. Since the oxidation process is fully microbial, an optimum amount of moisture has to be maintained continuously in the soil. The soil should not be leached until sufficient time has been allowed for most of the sulphur to oxidise. Limestone is a low cost amendment but the solubility is affected by pH of the soil and particle size of the amendment. Pressmud, a byproduct from sugar factories, contains about 70 to 80 % $CaCO_3$ and about 80 % organic matter. Since Ca is present as $CaCO_3$, it is slow acting amendment requiring acid or acid formers. As against carbonation process, pressmud from sugar factories employing sulphitation process has superior reclamation value, as it contains sulphate of lime instead of its carbonate. Like S, pyrite has to be oxidized first which is a slow process and the rate of reaction depends on particle size. Again the application of pyrites at higher rate markedly decreases its oxidation rate. It is a cheap amendment.

Gypsum is the most common amendment used for reclaiming saline-sodic as well as non-saline sodic soils. It is a low cost amendment and the rate of reaction in replacing Na is limited on its solubility in water, which is about 0.25 % at ordinary temperature. Khosla and Abrol have recommended a size of 0.59 mm (30 mesh) for maximum reactivity of gypsum. However, for practical purposes 2 mm size (10 mesh screen) is sufficient. While applying gypsum, mixing it in shallow depth (upper 10 cm depth) is more effective. It is applied by broadcast method or incorporated by disc plough. Gypsum is applied at the time of ponding or leaching. Gypsum directly prevents crust formation, swelling, dispersion and acts as mulch in case of surface application and indirectly increases porosity, structural stability, infiltration and hydraulic properties, soil tilth, drainage and leaching and reduces dry soil strength.

C. Chemical reactions of amendments in soil

The following chemical reactions illustrate the manner in which various amendments react in the different classes of alkali soils. In these equations the letter **X** represents the soil exchange complex.

GYPSUM: $2NaX + CaSO_4 \leftrightarrow CaX_2 + Na_2SO_4$

SULPHUR:
(1) $2S + 3O_2 \leftrightarrow 2SO_3$ (microbiological oxidation)
(2) $SO_3 + H_2O \leftrightarrow H_2SO_4$
(3) $H_2SO_4 + CaCO_3 \leftrightarrow CaSO_4 + CO_2 + H_2O*$
(4) $2NaX + CaSO_4 \leftrightarrow CaX_2 + Na_2SO_4$

LIME-SULPHUR (CALCIUM POLYPHOSPHATE):

(1) $CaS_5 + 8O_2 + 4H_2O \leftrightarrow CaSO_4 + 4H_2SO_4$
(2) $H_2SO_4 + CaCO_3 \leftrightarrow CaSO_4 + CO_2 + H_2O^*$
(3) $2NaX + CaSO_4 \leftrightarrow CaX_2 + Na_2SO_4$

IRON SULPHATE:

(1) $FeSO_4 + H_2O \leftrightarrow H_2SO_4 + FeO$
(2) $H_2SO_4 + CaCO_3 \leftrightarrow CaSO_4 + CO_2 + H_2O^*$
(3) $2NaX + CaSO_4 \leftrightarrow CaX_2 + Na_2SO_4$

* T e reaction of H_2SO_4 and $CaCO_3$ may also be written as follows : $H_2SO_4 + 2CaCO_3 \leftrightarrow CaSO_4 + Ca(HCO_3)_2$. Under these conditions the $Ca(HCO_3)_2$ as well as the $CaSO_4$ would be available for reaction with exchangeable sodium and 1 atom of sulphur when oxidized to H_2SO_4, could theoretically result in the replacement of 4 sodium ions by calcium.

D. Quantity of amendments to be added

The weight of the gypsum worked out on the basis of Schoonover's method is not entirely applied in the soil or 100 % of GR is never applied in the soil. These are evidences to show that even 50 % of the theoretical gypsum requirement for replacement of exchangeable Na in alkali soils has improved their physical properties and assisted response to management practices. Generally, 50 to 75 % of GR (as determined by Schoonover's method) has been found most satisfactory in many types of soils.

The equivalent proportion of different amendments in relation to 1 ton of gypsum is as follows :

Amendment	Weight in tones equivalent to 1 tone gypsum
Gypsum	1.000
Sulphuric acid	0.570
Sulphur	0.186
FeSO$_4$.7H$_2$O	1.620
Aluminium sulphate	1.290
Limestone (CaCO$_3$)	0.580
Lime sulphur (24 % S)	0.756

Among all the amendments, gypsum is the most common amendment that is used for the purpose of reclamation. The rate of addition of gypsum can be determined by estimating the gypsum requirement (GR) of a soil by Schoonover's method. Alternatively, the GR can also be determined by knowing the exchangeable Na in soil and working out the extent of reduction of Na on equivalent basis. The gypsum requirement for replacing 1 me of Na upto a soil depth of 15 cm comes to about 1.92 tonnes/ha. Since an ES { of 10 and below is considering safe for tolerable physical condition of the soil, replacement by calcium to this level is all that is attempted in practice.

E. The organic amendments

The organic amendments as such do not help in replacing the exchangeable Na as against the gypsum or other amendments. Primarily, they improve the physical condition of the soil by improving the aggregation in the soil. The most common organic amendment is the FYM which is added in the first year of reclamation @ 50 tonnes/ha and is reduced to half in succeeding years. The efficiency of gypsum has been found to increase when it is applied along with FYM. Molasses and pressmud, which are sugar factory waste, have also been used. Molasses contain 60 – 70 % carbohydrates, 4 – 5 % potash, 2 % lime and 0.5 % each of N, P_2O_5, H_2SO_4 and iron and aluminium oxide.

Green manuring with Dhaincha (*Sesbania aculeata*) has been found most successful. The juice of green plants can neutralize high alkalinity, its initial pH being 4.01, with only slight rise even within a month. In black cotton soil, it thrives well under moderately saline conditions and can with stand high alkalinity, water logging or drought so that it is remarkably suited in that region to alkali soils, characterized by such adverse conditions. Sulphurated hydrogen is generated by the decomposition of Dhaincha.

Paddy straw or rice husk have also been used at a rate varying between 15 to 30 tonnes/ha. Weeds like *Argemone mexicana* has been found very suitable for alkali soils. It contains (on dry weight basis) 1.8 % KNO_3, 0.3 % $CaHPO_4$, 0.4 % $CaSO_4$, 4.2 % organic acid and 0.8 % sugar. When powder of argemone was added to the soil @ 2.5 tonnes/ha, it lowered the soil pH from 10.0 to 7.8 which slowly leveled to 8.5 in 30 days. The other weeds found suitable for the purpose of green manuring are *Ipomea grandiflora* and *Pongamia glabra*. The Russian workers have suggested the addition of cellulose with sufficient addition of nitrogen for easy decomposition.

Management practices

During the process of reclamation of salt affected soils the field is not kept fallow. Growing of crop is always practiced as the roots of the growing crop exert a marked beneficial effect on the process of leaching by improving the state of aggregation of soil. A crop can be grown by adopting following practices :

A. Selection of crop: The selection of crop is based on tolerance of a crop to either salinity or sodicity. The list of salt tolerant as well as sodium tolerant crops is given below :

Table 10.4.: Yield decrease of certain crops due to variable salt levels in soil solutions

Crop	EC x 10^3 of saturation extract causing yield decreases		
	10 %	25 %	50 %
Field crops			
Barley for grain	12	16	18
Sugar beet	10	13	16
Cotton	10	12	16
Safflower	8	11	12
Wheat	7	10	14
Sorghum	6	9	12
Soybean	5.5	7	9
Sugarcane	3	7	5.5
Rice	5	6	8
Corn	5	6	7
Bean	1.5	2	2.5
Vegetable crops			
Beet (garden)	8	10	12
Tomato	4	6.5	8
Cabbage	2.6	4	7
Potato	2.5	4	6
Sweet potato	2.5	3.5	6
Onion	2	2.5	4
Carrot	1.5	2.5	4

The salt tolerance limits given in Table 1 are general. The salt tolerance of a crop would vary according to age of the crop or the growth stage. It will also vary according to variety of a given crop. Hence considerable work is needed for establishing these limits for local crops and their varieties.

PROBLEMATIC SOILS

The tolerance of various crops to ESP is given in Table 2.

Table 10.5.: Tolerance of various crops to ESP

ESP	Class	Crop
2 - 10	Very sensitive	Deciduous fruits, Nuts Citrus, Avocado
10 - 20	Sensitive	Beans
20 - 40	Moderately tolerant	Clover, Oats, Tall fescue, Rice, Dallis grass
40 - 60	Tolerant	Wheat, Cotton, Alfalfa Barley, Tomatoes, Beet
More than 60	Highly tolerant	Crested wheat grass
		Fairway wheat grass
		Tall wheat grass
		Rhodes grass

Table 10.6: Screening of varieties of different crops for saline condition

Crop	Varieties
Wheat	: Kharachia, J-24, Popatia, Arnej-206, Sonalika, Kalyan Sona
Bajra	: B.K.-560, GHB-235, MH-169, MH-179, GHB-227 for fodder
Cotton	: G.Cot.D.H.-7, Dhumal, Kalyan
Sugarcane	: Co.8338, Co.791
Groundnut	: JL-24, J-11, Robert, Punjab-1
Castor	: GCH-4, SKF-73, GAUCH-1, VP-1
Jowar	: Gundari, C-10-2, CSH-5, S.R.F.-204 for fodder
Paddy	: T.N.-1, Jaya
Mustard	: Varuna, A.S.-10
Sunflower	: EC-68414, EC-68415
Pigeon pea	: GT-100, GT-1
Chick pea	: ICCC-4, JCP-29

B. Tillage: Deep tillage should be practiced as it increases the permeability of the soil thereby facilitating leaching of salts. It also makes the root zone more friable. If the subsoil layer has accumulation of salts or sodium, the deep tillage is not advisable.

C. Layout: Crop sown in furrow show better performance than those in flat layout. Salt accumulate on ridges leaving the furrow relatively free of salts. However in alternate furrow irrigation system, sowing at ridges can be advocated.

D. Seed rate and spacing: Higher seed rate and closer spacing have been found satisfactory as high plant populations insures against the failure of germination due to salt stress.

E. Irrigation and drainage: Maintenance of low moisture stress by frequent irrigation and applying water in excess enables the plant to grow better. Among the methods of irrigation, the minimum salt accumulation takes place in the order of drip > sprinkler > flood basin > ridge furrow.

F. Fertilizers: A higher dose of N, P_2O_5 and K_2O than recommended dose has given higher yields of many crops such as rice, barley, wheat, etc. Addition of micronutrients, particularly Zn, Fe and Mn has helped in increasing the yield. In saline conditions urea, single super phosphate and calcium ammonium nitrate, while in sodic conditions ammonium sulphate and diammonium phosphate found more effective. In highly sodic condition, foliar application of urea is the only effective and economical method of fertilization. In highly saline condition phosphatic fertilizer did not found economical.

Important formula used in calculating examples

(A) Water sample

- % soluble salt = EC x 640, where EC in dS/m
- ppm salt = % salt x 10,000
- % soluble salt = ppm salt/10,000
- RSC = $(CO_3 + HCO_3) - (Ca + Mg)$ in me/l
- Total cation concentration in me/l = EC x 10 where EC in dS/m
- Na me/l = EC (dS/m) x 10 − (Ca + Mg)

- $SAR = \sqrt{\dfrac{Na^+}{\dfrac{Ca^{2+}Mg^{2+}}{2}}}$

Volume of water for 5 cm hectare water = $\dfrac{100 \times 100 \times 5}{100} = 500$ cu.m.

1 cu.m. water = 1000 kg or litre

kg salt added due to 5 ha = $\dfrac{Cu.m. \times 1000 \times \% \text{ salt}}{100}$

(12) ESP = $\dfrac{Na}{Total\ cations} \times 100$

Some useful conversion factors

Note: The SI unit of conductivity is 'Siemens' symbol 'S' per metre. The equivalent non-SI unit is 'mho' and 1 mho = 1 Siemens. Thus for those unused to the SI system mmhos/cm can be read for dS/m without any numerical change.

Conductivity 1 S cm^{-1} (1 mho/cm) = 1000 mS/cm (1000 mmhos/cm)

1 mS/cm^{-1} (1 mmho/cm) = 1 dS/m = 1000 mS/cm (1000 micromhos/cm)

Conductivity to mmol (+) per litre:

mmol (+)/l = 10 × EC (EC in dS/m)

for irrigation water and soil extracts in the range 0.1-5 dS/m.

Conductivity to osmotic pressure in bars:

OP = 0.36 × EC (EC in dS/m)

for soil extracts in the range of 3-30 dS/m.

Conductivity to mg/l:

mg/l = 0.64 × EC x 10^3, or (EC in dS/m)

mg/l = 640 × EC

for waters and soil extracts having conductivity up to 5 dS/m.

mmol/l (chemical analysis) to mg/l:

Multiply mmol/l for each ion by its molar weight and obtain the sum

10.2. ACID SOILS

A. Causes of soil acidity

1. *Excessive rainfall*: In soils of dry region, a large supply of bases is usually present because little water passes through the soil. With an increase in rainfall, the content of soluble salts is reduced to a low level and gypsum and CaCO$_3$ are removed in the order named. With further increase in rainfall, a point is reached at which the rate of removal of bases exceeds the rate of liberation from non-exchangeable forms. The considerable loss of bases due to intensive rainfall and leaching reduces the pH of the soil as well as increase the concentration of H$^+$ on exchange complex.

2. *Ionization of water*: The water may ionize and contribute H$^+$ on exchange complex as follows : H$_2$O → HOH → H$^+$ OH$^-$ → H$^+$[X] + Bases + OH$^-$

3. *Contact exchange*: The contact exchange between exchangeable H on root surface and the bases in exchangeable form on soil particle may take as follows

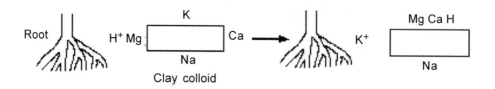

4. **Soluble acid production** : The decomposition of organic matter in the soil produces many organic as well as inorganic acids. These acids may contribute H on exchange complex.

5. **Use of nitrogenous fertilizers** : Continuous use of nitrogenous fertilizers containing NH_4-N or giving NH_4-N on hydrolysis (i.e. urea) produce various acids in soils e.g. 1 mole of NH_4 in NH_4NO_3 gives 2 moles of HNO_3; 1 mole of $(NH_4)_2SO_4$ gives 2 moles of HNO_3 + 1 mole of H_2SO_4; 1 mole of NH_4OH gives 1 mole of HNO_3. Thus, continuous use of such fertilizers will produce acidity in soil.

6. **Oxidation of FeS** : FeS or iron poly sulphide accumulates under anaerobic conditions as a result of reduction of Fe^{3+} and SO_4. Under aerobic conditions, they will be oxidized and will produce H_2SO_4. Under such conditions, soil pH values of 2 to 4 are frequently observed.

$$4FeS_2 + 15O_2 + 2H_2O \rightarrow 2Fe_2(SO_4)_3 + 2H_2SO_4$$

7. **Hydrolysis of Fe^{3+} and Al^{3+}**: The Fe^{3+} and Al^{3+} ions may combine with water and release H^+ as follows : $Al + H_2O \rightarrow Al(OH) + H$

$$Al(OH) + H_2O \rightarrow Al(OH)_2 + H$$
$$Al(OH)_2 + H_2O \rightarrow Al(OH)_3 + H$$

The hydrogen produced may enter on exchange complex.

8. **Acidic parent material**: Some soils have developed from parent materials which are acid, such as granite and that may contribute to some extent soil acidity.

9. **Acidification from the air** : Industrial exhausts, if contain appreciable amount of SO_2 may cause acidity in soil in course of time due to dissolution of SO_2 in water (rain) as follows :

$$SO_2 + H_2O \rightarrow H_2SO_3$$
(Rain water)
$$2H_2SO_3 + O_2 \rightarrow 2H_2SO_4$$
(Sulphuric acid)

Table 10.7. Extent of acid soils in different state (m ha)

Sr.No	States	pH<5.5	pH 5.5-6.5	Total
1	A.P	–	0.40	0.40
2	Assam	2.33	2.33	4.66
3	Bihar	0.04	2.32	2.36
4	Chhattisgadh	6.45	4.39	1.84
5	Jharkhand	1.0	5.77	6.77
6	Karnataka	0.06	3.25	3.31
7	Kerala	3.01	0.75	3.76
8	M.P.	1.12	10.60	11.72
9	Maharashtra	0.21	4.33	4.51
10	Orissa	0.26	8.41	8.67
11	Tamil Nadu	0.56	4.29	4.85
12	West Bengal	5.6	1.2	4.76
	TOTAL	15.6	51.04	66.64

B. Forms of Acidity

Because of the increase in H ion concentration in soil solution some of them occupy the position on exchange complex because of its very high replacing ability. The situation gives high H ion concentration in soil solution and on exchange complex, or the acidity in solution and on exchange complex. The acidity in soil solution is known as Active Acidity and is measured by pH. The acidity on exchange complex is known as Passive Acidity or Reserve Acidity and is measured by determining the exchangeable H by $BaCl_2$ + Triethanol amine reagents. The total acidity or the titratable acidity is summation of H ion concentration present in solution as well as on exchange complex which can be measured by titration. All the forms of acidity are in equilibrium.

C. Problems in acidic soils

Problems of soil acidity may be divided into three groups:
1. Toxic effects
    (a) Acid toxicity
    (b) Toxicity of elements

2. Nutrient availability
   (a) Exchangeable bases
   (b) Nutriment imbalances
3. Microbial activity

1. Toxic effects

(a) *Acid Toxicity*: The higher hydrogen ion concentration is toxic to plants under strong acid conditions of soil. The acid toxicity includes possible toxicities of acid anions as well as H⁺ ions.

(b) *Toxicity of elements*

(i) *Iron, Manganese and Aluminium*: The concentration of these ions (Fe^{2+}, Mn^{2+} and Al^{3+}) in soil increased in acidic condition to a very high and toxicity of these elements develop.

2. Nutrient Availability

(i) *Exchangeable Bases*: There are two aspects of availability of exchangeable bases i.e., ion uptake process and the release of bases from the exchangeable form may be adversely affected due to soil acidity. Deficiency of bases like Ca^{2+} and Mg^{2+} are found in acid soils.

(ii) *Nutrient Imbalances*: It is evident that soluble iron, aluminium and manganese are usually present in their higher concentrations under moderate to strong acid soils. Phosphorus reacts with these ions and produces insoluble phosphatic compounds rendering phosphorus unavailable to plants. Besides these, fixation of phosphorus by hydrous oxides of iron and aluminium or by adsorption, the availability of phosphorus is decreased. In acid soils, iron, manganese, copper and zinc are abundant, but molybdenum is very limited and unavailable to plants. In acid soils having very low pH, the availability of boron may also be decreased due to adsorption on sesquioxides, iron and aluminium hydroxy compounds. Nitrogen, K_2O and sulphur become less available in an acid soil having pH less than 5.5.

3. Microbial Activity: It is well-known that soil organisms are influenced by fluctuations in the soil reaction. Bacteria and *actinomycetes* function better in soils having moderate to high pH values. They can not show their activity when the soil pH drops below 5.5. Nitrogen fixation in acid soils is greatly affected by lowering the activity of *Azotobacter* sp. Besides these, soil acidity also inhibits the symbiotic nitrogen fixation by affecting the activity of *Rhizobium* sp. Fungi can grow well under very acid soils and caused various diseases like root rot of tobacco, blights of potato, etc.

E. General characteristics of Acid Soils

1. Acid soils have low pH and high proportion of exchangeable H^+ and Al^{3+}.
2. Kaolinite and illite types of clay minerals are dominant in these soils.
3. These soils have low CEC and low base saturation.
4. These soils have high toxic concentration of Al, Fe and Mn and deficiency of Ca and Mg.
5. These soils have nutrients and microbial imbalances.
6. These soils are low in available P2O5. Soil acidity inhibits biological N-fixation.

F. Reclamation of Acidic Soils

Principles of Liming Reactions: The reclamation of acidic soils is done by addition of liming material which may be calcitic limestone ($CaCO_3$) or dolomitic limestone [$CaMg(CO_3)_2$]. The rate of lime requirement is determined in the laboratory by method of Shoemaker (1961). The particle size of liming material affects the rate of neutralization reaction. Both these lime stones are sparingly soluble in pure water but do become soluble in water containing CO_2. The greater the partial pressure of CO_2 in the system, the more soluble the limestone becomes. Dolomite is somewhat less soluble than calcite. The reaction of limestone ($CaCO_3$) can be written as:

$$CaCO_3 + H_2O + CO_2 \rightarrow Ca(HCO_3)_2$$
$$Ca(HCO_3)_2 \rightarrow Ca^{2+} \downarrow + 2HCO_3^-$$

(Takes part in cation exchange reactions)

$$H^+ + CO_3^- \rightarrow H_2CO_3 \leftrightarrow H_2O + CO_2$$

(From soil solution) (from lime)

In this way hydrogen ions (H^+) in the soil solution react to form weakly dissociated water, and the calcium (Ca^{2+}) ion from lime stones is left to undergo cation exchange reactions. The acidity of the soil is, therefore, neutralized and the per cent base saturation of the colloidal material is increased.

Why Gypsum is not considered as a Liming Material?

Gypsum is not considered as liming materials because on its application to an acid it dissociates into (Ca^{2+}) and sulphate (SO_4^{2-}) ions:

$$CaSO_4 \leftrightarrow Ca^{2+} + SO_4^{2-}$$

SOIL FERTILITY AND NUTRIENT MANAGEMENT

The accompanying anion is sulphate and it reacts with soil moisture produces mineral acid (H_2SO_4) which also increases soil acidity instead of reducing soil acidity.

G. Management of Acid Soils

(a) Ameliorating the soils through the application of amendments

(i) Oxides of lime (CaO) : When oxides of lime like CaO is applied to an acid soil, it reacts almost immediately as follows :

$$\text{Soil}\begin{matrix}H^+\\ \\Al^{3+}\end{matrix} + H_2O + 2CaO \rightarrow \text{Soil}\begin{matrix}Ca^{2+}\\ \\Ca^{2+}\end{matrix} + Al(OH)_3$$

(ii) Hydroxides of lime [Ca(OH)$_2$] : When hydroxides of lime like $Ca(OH)_2$ is applied for the reclamation of an acid soil, the following chemical reaction takes place :

$$\text{Soil}\begin{matrix}H^+\\ \\Al^{3+}\end{matrix} + 2Ca(OH)_2 \rightarrow \text{Soil}\begin{matrix}Ca^{2+}\\ \\Ca^{2+}\end{matrix} + Al(OH)_3 + H_2O$$

(iii) Carbonates of lime (CaCO$_3$): When carbonates of lime like calcite is applied to an acid soil, a part of $CaCO_3$ undergo solution and combines with H_2CO_3 to form soluble $Ca(HCO_3)_2$. This calcium bicarbonate in solution from reacts with the soil colloids with the evolution of CO_2 as follows :

$$CaCO_3 + H_2CO_3 \rightarrow Ca(HCO_3)_2$$

$$\text{Soil}\begin{matrix}H^+\\ \\H^+\end{matrix} + 2Ca(HCO_3)_2 \rightarrow \text{Soil}\begin{matrix}Ca^{2+}\\ \\Ca^{2+}\end{matrix} + H_2O + 2CO_2$$

And the rest portion of the limestone goes in close contact with the soil colloids in solid condition as follows:

$$\text{Soil}\begin{matrix}H^+\\ \\H^+\end{matrix} + CaCO_3 \rightarrow \text{Soil}\begin{matrix}Ca^{2+}\\ \\ \end{matrix} + H_2O + CO_2$$

Beneficial effect of lime

(1) Lime makes P_2O_5 more available.
(2) Lime increase availability of N, increase nitrification and N- fixation.

(3) Increase soil pH favours the microbial activity and increase organic matter decomposition and nutrient transformation for root growth.

(4) Mo an essential element to *Rhizobium* in N fixation process increases with increase in soil pH following lime.

(5) Reduce toxicity of Al, Fe and Mn.

(6) Lime is essential source of essential Ca as well as Mg if dolomitic lime stone has been applied as liming material.

(7) It causes an increase in CEC, which reduces the leaching of base cations, particularly K.

(b) **Selection of crop:** The crop should be selected on the basis of their tolerance to acidity. The relative yield of different crops at different pH values is given in following table.

Crops	Relative yield at pH value				
	4.7	5.0	5.7	6.8	7.5
Corn	34	73	83	100	85
Wheat	68	76	89	100	99
Oats	77	93	99	98	100
Barley	0	23	80	95	100
Alfalfa	2	9	42	100	100
Soybeans	65	79	80	100	93

High acid tolerant crops	:	Rice, potato, sweet potato, oat, castor, .
Moderate acid tolerant crops	:	Barley, wheat, maize, turnip, brinjal, *etc.*
Slightly acid tolerant crops	:	Tomato, carrot, red clover, *etc.*
Al tolerant crop		
Agricultural crops	:	Pineapple, Coffee, Tea, Rubber, Cassava, Sweet potato, Rice, Finger millet

(c) **Use of basic fertilizers:** $NaNO_3$ and basic slag, etc.

(d) **Soil and water management**: Proper soil and water management checks leaching of bases and enhances decomposition of organic matter.

10.3. ACID SULPHATE SOILS

Soil with sufficient sulphides (FeS_2 and others) to become strongly acidic when drained and aerated enough for cultivation are termed acid sulphate soils or as the Dutch refer to those soils *cat clays*. When allowed to develop acidity, these soils are usually more acidic than pH 4.0. Before drainage, such soils may have normal soil pH and are only *potential acid sulphate soils.* Generally acid sulphate soils are found in coastal areas, where the land is inundated by salt water. In India, acid sulphate soil is mostly found in Kerala, Orissa, Andhra Pradesh, Tamil Nadu and West Bengal. The area covered under acid sulphate soils in Thailand and India combinely is about 2 million acres.

Formation of Acid Sulphate Soils

Lands inundated with waters that contain sulphates, particularly salt waters, accumulate sulphur compounds, which in poorly aerated soils are bacterially reduced to sulphides. Such soils are not usually very acidic when first drained in water.

When the soil is drained and then aerated, the sulphide (S^{2-}) is oxidized to sulphate (SO_4^{2-}) by a combination of chemical and bacterial actions, forming sulphuric acid (H_2SO_4). The magnitude of acid development depends on the amount of sulphide present in the soil and the conditions and time of oxidation. If iron pyrite (FeS_2) is present, the oxidized iron accentuates the acidity but not as much as aluminium in normal acid soils because the iron oxides are less soluble than aluminium oxides and so hydrolyze less.

Reactions Involving the Formation of Acid Sulphate Soils

Acid sulphate soils form due to oxidation of sulphides in soils. The slow oxidation of mineral sulphides in soils is non-biological until soil pH reaches an acidity of pH 4.0, but the process is not well understood. Below pH 4.0, the bacteria *Thiobacillus ferroxidans* are the most active oxidizers and the activity builds up rapidly.

Microbial oxidation

$$2S + 3O_2 + 2H_2O \longrightarrow H_2SO_4$$
(Elemental sulphur) (Sulphuric acid)

Microbial oxidation

$$H_2S + 2O_2 \longrightarrow H_2SO_4$$
(Hydrogen sulphide) (Sulphuric acid)

Non-biological

$$2FeS_2 + 2H_2O + 7O_2 \longrightarrow 2FeSO_4 + 2H_2SO_4$$
(Iron pyrite) (Ferrous sulphate) (Sulphuric acid)

Accelerated by bacteria *(Thiobacillus ferroxidans):*

$$4FeSO_4 + O_2 + 2H_2SO_4 \longrightarrow 2Fe_2(SO_4)_3 + 2H_2O$$
(Ferrous sulphate) (Ferric sulphate)

Rapid in acid pH (non-biological):

$$FeS_2 + 7Fe_2(SO_4)_3 + 8H_2O \longrightarrow 15FeSO_4 + 8H_2SO_4$$
(Ferrous sulphate)

Management of Acid Sulphate Soils

Management techniques are extremely variable and depend on many specific factors viz., the extent of acid formation, the thickness of the sulphide layer, possibilities of leaching or draining the land, etc. The general approaches for reclamation are suggested below:

(i) Keeping the area flooded: Maintaining the reduced condition of flooded (anaerobic) soil inhibits acid development, which requires oxidation. This solution almost limits the use of the urea to rice growing. Unfortunately, droughts occur and can in short time periods cause acidification of these soils. The water used to flood the potential acid sulphate soils often develop acidity and injure crops.

(ii) Controlling water table: If a non-acidifying layer covers the sulphuric horizon, drainage to keep only the sulphuric layer under water (anaerobic) is possible.

(iii) Liming and leaching: Liming is the primary way to reclaim any type of acid soil. Acid soil may require 11-45 MT per hectare of lime in a 20-year period whereas, acid sulphate soils may require from several metric tons per hectare per year, up to 224 MT per hectare (100 t/acre) within a 10 year period or less.

If these soils are leached during early years of acidification, lime requirements are lowered. Leaching, however, is difficult because of the high water table commonly found in this type of soil and low permeability of the clay. Sea water is sometimes available for preliminary leaching.

10.4. CALCAREOUS SOILS

In the context of agricultural problem soils, calcareous soils are soils in which a high amount of calcium carbonate dominates the problems related to agricultural land use. They are characterized by the presence of calcium carbonate in the parent material and by a calcic horizon, a layer of secondary accumulation of carbonates (usually Ca or Mg) in excess of 15% calcium carbonate equivalent and at least 5% more carbonate than an underlying layer.

Calcareous soils cover more than 30% of the earth surface, and their $CaCO_3$ content varies from a few percent to 95%. Hagin and Tucker (1982) define calcareous soil as a soil that its extractable Ca and Mg levels exceed the cation exchange capacity.

Calcareous soil divided into following 4 classes:

Group	Calcium Carbonate(%)
(i) Slight calcareous	0-5
(ii) Moderate calcareous	5-10
(iii) Strong calcareous	10-20
(iv) Very strong calcareous	20-30

Origin of Calcareous Soils

1. Calcareous soils occur naturally in arid and semi-arid regions because of relatively little leaching

2. They also occur in humid and semiarid zones if their parent material is rich in $CaCO_3$, such as limestone, shells or calcareous glacial tills, and the parent material is relatively young and has undergone little weathering.

3. Some soils that develop from calcareous parent materials can be calcareous throughout their profile. This will generally occur in the arid regions where precipitation is scarce. In other soils, $CaCO_3$ has been leached from the upper horizons, and accumulated in B or C horizons. These lower $CaCO_3$ layers can be brought to the surface after deep soil cultivation.

4. In some soils, the $CaCO_3$ deposits are concentrated into layers that may be very hard and impermeable to water. These *caliche* layers are formed by rainfall leaching the salts to a particular depth in the soil at which water content is so low that carbonates precipitate.

5. Soils can also become calcareous through long period of irrigation with water containing dissolved $CaCO_3$

Main production constraints

1. Calcareous soils develop in regions of low rainfall and must be irrigated to be productive. Therefore one of the main production constraints is the availability of water for irrigation.

2. The quality of the irrigation water is of crucial importance for sustainable agricultural production on calcareous soils. Frequently, the irrigation water is the cause of many management problems. Almost all waters used for irrigation

contain inorganic salts in solution. These salts may accumulate within the soil profile to such concentrations that they modify the soil structure, decrease the soil permeability to water, and seriously injure plant growth.

3. Crusting of the surface may affect not only infiltration and soil aeration but also the emergence of seedlings.
4. Cemented conditions of the subsoil layers may hamper root development and water movement characteristics.
5. Calcareous soils tend to be low in organic matter and available nitrogen.
6. The high pH level results in unavailability of phosphate (formation of unavailable calcium phosphates as apatite) and sometimes reduced micronutrient availability e.g. zinc and iron (lime induced chlorosis). There may be also problems of potassium and magnesium nutrition as a result of the nutritional imbalance between these elements and calcium.

Role of $CaCO_3$ in plant nutrition: The carbonates are characterized by a relatively high solubility, reactivity and alkaline nature; their dissolution resulting in a high solution HCO_3^- concentration which buffers the soil in the pH range of 7.5 to 8.5:

$$CaCO_3 + H_2O \rightarrow Ca^{2+} + HCO_3^- + OH^-$$

Symptoms of impaired nutrition in calcareous soils are chlorosis and stunted growth. This is attributed to the high pH and reduced nutrient availability, as direct toxicity of bicarbonate ions (HCO_3^-) to physiological and biochemical systems are much less likely. The availability of P and Mo is reduced by the high levels of Ca and Mg that are associated with carbonates. In addition, Fe, B, Zn, and Mn deficiencies are common in soils that have a high $CaCO_3$ due to reduced solubility at alkaline pH values

Nitrogen Management in Calcareous Soils: Ammonium fertilizers are superior as compared to nitrate fertilizers in very slightly alkaline soils (pH 7-7.5) due to its side effect as a soil acidifier ($2NH_4^+ + 3O_2 \Leftrightarrow 2NO_3^- + 8H^+$). When ammonium fertilizers are surface applied to calcareous soils, the following reaction occurs:

$$2NH_4A + CaCO_3 \Leftrightarrow (NH_4)_2CO_3 + CaA_2$$

Where, A represents the accompanying anion in the ammonium fertilizer. The $(NH_4)_2CO_3$ product is unstable and decomposes as follows:

$$(NH_4)_2CO_3 + H_2O \Leftrightarrow 2NH_3 + H_2O + CO_2$$
$$NH_3 + H_2O \Leftrightarrow NH_4^+ + OH^-$$

Gaseous NH_3 is formed and lost by diffusion into the atmosphere. If the CaA salt is an insoluble one, then the reaction will proceed to the right causing more $(NH_4)_2CO_3$ to be formed and thus more NH_3 is generated and volatilized. But when the accompanying anion forms a soluble Ca compound, less $(NH_4)_2CO_3$ will be formed. Therefore those sources which form precipitates of low solubility with Ca such as ammonium sulfate and phosphate will suffer larger ammonia losses than ammonium nitrate or chloride, which forms soluble reaction products with Ca. Regarding NH_3 losses from urea, additional practices include mixing the urea with KCl, $CaCl_2$ or TSP, and the use of granular forms, urease inhibitors, and sulfur-coated urea.

Phosphorus Management in Calcareous Soils: Phosphorus availability in calcareous soils is usually restricted. At higher pH values, phosphate anions react with Ca and Mg to form phosphate compounds of limited solubility. Added phosphate is precipitated as dicalcium phosphate dihydrate (DCP; $CaHPO_4.2H_2O$) or octacalcium phosphate (OCP; $Ca_8H_2[PO_4]_6.5H_2O$). Hydrolysis of DCP to OCP increases with increasing pH of soil. OCP is converted over a long period of time to less soluble apatites (hydroxyapatite, $Ca_{10}[OH]_2[PO_4]_6$) and fluorapatite ($Ca_{10}F_2[PO_4]_6$). Soluble P fertilizers (triple super phosphate, ammonium phosphates) are the preferred source in calcareous soils. Application of P in bands instead of broadcasting and mixing with a large soil volume, and as large granules instead of a fine powder - decreases the reversion to less soluble forms by reducing the contact between fertilizer and soil

Potassium and Magnesium Management in Calcareous Soils

Available K and Mg are usually found in an adequate supply in calcareous soils. However, an imbalance between plant available Mg, Ca and K ions may lead to Mg and/or K deficiencies to crops. High Ca levels in soils suppress Mg and K uptake by crops in part, presumably, through the competition between Ca, Mg and K Therefore, crops growing on soils high in Ca often require above normal levels of Mg and K fertilization for satisfactory nutrition. The K supplying power of calcareous soils was found lower than that of alluvial soils, even when their content of non-exchangeable K was similar. In cases where soil-applied fertilizer is ineffective, the only way to increase leaf Mg or K concentration may be through foliar application of water-soluble fertilizers, such as magnesium nitrate $[Mg(NO_3)_2]$ or potassium nitrate (KNO_3).

Iron Management in Calcareous Soils: Calcareous soils may contain high levels of total Fe, but in forms unavailable to plants. Visible Fe deficiency, or Fe chlorosis, is common in many crops. However, owing to the nature and causes of Fe chlorosis, leaf Fe concentration is not necessarily related to degree of chlorosis; in chlorotic plants Fe concentrations can be higher or lower than those in normal plants. Thus, this disorder on calcareous soils is not always attributable to Fe deficiency; this condition is known as *lime-induced Fe chlorosis.*

The primary factor associated with Fe chlorosis under calcareous conditions appears to be the effect of the bicarbonate ion (HCO_3) in reducing Fe uptake and translocation to the leaves. In some other cases, the lime-induced chlorosis is related to a high Fe level in the chlorotic leaves, which has to be somehow unavailable or immobilized inside the leaf tissue Inorganic sources of Fe such as ferrous sulfate ($FeSO_4$) or ferric sulfate [$Fe_2(SO_4)_3$] have a very limited effect unless applied very frequently at extremely high rates.

Chelate: Iron chlorosis can be corrected through soil application of Fe chelates. Plants can take up the soluble chelate as complete molecules and then metabolize the metal.

Potassium – Iron and Ammonium-Iron interaction: Potassium fertilization at rates of 135 to 405 mg K/kg soil ameliorated iron chlorosis in groundnut grown in an extremely calcareous soil (63% $CaCO_3$). Supplying ammonium sulfate in the presence of nitrification inhibitor (nitrapyrin) reduced Fe chlorosis in groundnut grown on very high calcareous soil (98% $CaCO_3$) in Israel. These results are attributed to the cation-anion balance of ion uptake: the plant takes up more cations than anions, there is an efflux of H^+ to correct this imbalance, the rhizosphere is acidified and consequently iron is more available to roots.

Organic Manures and Sewage sludge: Applications of manure can help to correct Fe chlorosis.

Foliar fertilization: In citrus, foliar application of ferrous sulfate ($FeSO_4$), ferrous sulfate heptahydrate ($FeSO_4 \cdot 7H_2O$) or Fe chelates has not proven effective because of poor Fe translocation within the leaf. The use of foliar sprays also increases the possibility of fruit and/or leaf burn.

Zinc and Manganese Management in Calcareous Soils: Zinc and Mn deficiencies are clearly pH-dependent, and both Zn and Mn concentration in solution decreases 100-fold for each unit increase in pH.

The most common inorganic Zn and Mn fertilizers are the sulfates ($ZnSO_4$, $MnSO_4$) and their oxide forms (ZnO, MnO). Broadcast application of these compounds to correct Zn or Mn deficiencies in calcareous soils is not effective; Zinc is also available in chelated forms, including Zn-EDTA and Zn-EDDHA. Chelated Zn when applied to calcareous soils remains soluble and available to plants considerably longer than the inorganic forms. However, soil application of chelated Zn is rarely economical. Manganese chelates have limited effectiveness in calcareous soils and are not normally used.

The least expensive way to apply Zn and Mn is through foliar sprays. In addition to the forms listed above, a number of other Zn and Mn formulations are available for foliar spraying, including nitrates and organically chelated forms. Foliar application of Zn has to be repeated several times per year, due to the limited translocation from older to new leaves.

Copper, Boron and Molybdenum Management in Calcareous Soils

Copper: Copper solubility is pH-dependent and it decreases with increasing pH. The most common source of Cu for soil and foliar application is copper sulfate ($CuSO_4 \cdot 5H_2O$) and also Cu chelates.

Boron: Soil pH affects B availability more by sorption reactions than by formation of less soluble compounds. Availability of B is highest in the pH range of 5.5-7.5. There is an interaction between B availability and the presence of Ca ions. High levels of Ca at high pH reduce the uptake of B. This may explain the fact that high B levels in calcareous soils, considered as toxic in other conditions, do not produce B toxicity in crops.

Molybdenum: Mo deficiencies are not known in calcareous soils, as Mo availability increases with pH. Molybdenum toxicity to plants has not been reported. However, high Mo levels can accumulate in alkaline soils, which may be toxic to livestock this condition is known as *molybdenosis*

Acidifying calcareous soils: Acidification may be needed for crops with a low optimum pH range grown on calcareous soils. To decrease soil pH, the $CaCO_3$ would have to be dissolved or neutralized by adding acid or acid forming materials *viz.*, elemental S, Sulphuric acid, aluminum sulphate, ammonium polysulphide *etc.* In most field crop situations, reducing soil pH by neutralizing $CaCO_3$ is not practical, for example, the quantity of elemental S needed to neutralize a soil with only 2% $CaCO_3$ (0-15 cm depth) is estimated by:

2% $CaCO_3$ = 2 g $CaCO_3$/100 g = 2000 mg/100g = 40 me/100g

40 me $CaCO_3$ required 40 me S to neutralize

= 40 me S/100g = 640 mg/100g = 6400 mg/1000g

= 6400 x 2.24 = 14336 kg S/ha = 14.3 t S/ha

Once neutralized, soil pH would likely to about the same as before neutralization because the CEC would still be nearly about 100% saturated with basic cation

(100% BS). To ultimately lower soil pH below 7, additional S would be needed to produce H^+ and Al^{3+}, which would in turn reduce the % BS necessary to lower soil pH.

Management of Calcareous Soil

Management and reclamation of calcarous soil are not difficult because pH in such soil is not very high. Generally, there is no need of chemical amendment for reclamation of calcareous soil. The calcareous soil can be managed in the following ways.

(i) **Tillage Operation.** Light (sandy) calcareous soil develops large number of pore spaces due to flocculation. These type of soils have poor water-holding capacity. Therefore, such type of soil needs compaction by plank or roller to increases the water-holding capacity.

(ii) **Application of Organic Manure.** When sufficient amount of farm yard manure, compost and green manure are added, the amount of carbon dioxide and acid increases and as a result pH of soil decrease.

(iii) **Use of Chemical Fertilizer.** Availability of phosphorus is low in calcareous soil. To increase the availability of P, the phosphatic fertilizers should be used in the following manner :

(a) Phosphatic fertilizers shoud be used near the roots of plant.

(b) Use of phosphatic fertilizers in ball form also increases its availability.

(c) May be used in split dose.

(iv) **Use of Micronutrients.** Addition of micronutrients like, zinc, copper, iron would be helpful in increasing yield.

10.5. CLAYEY/BLACK SOIL

The name *black* is given to soils that are very dark in colour and turn extremely hard on drying and sticky and plastic on wetting and hence are very difficult to cultivate and manage.

These soils are dark in colour, rich in clay content and have characteristics associated with shrink and swell properties. The high clay content (>30% up to atleast 50 cm of the soil surface), which in dry state, develop typical cracks and are 1 cm or more wide and reach a depth of 50 cm or more. Such soils are often called as heavy, cracking-clay soils.

Salient Characteristics

The soil that have develop on basalt, or alluvium derived from basaltic rocks are generally deep (100-150 cm) to very deep (>150 cm) and those developed on gneisses and schists are moderately shallow (50-75 cm) to moderately deep (75-100 cm). They have uniform colour throughout the pedon, except where they overlie saprolite (*murrum*). Other salient characteristics of these soils are outlined as under:

- These are highly clayey with clay content ranging from 30 to 80 per cent. The terminal infiltration rates of these soils are generally low, ranging from 0.2 to over 29 mm/hr. The clay type is typically montmorillonitic that has high degree of swelling on wetting and shrinkage on desiccation.
- Being rich in smectite clay minerals, they have high exchange capacity (30-60 cmol (P^+)/kg) and are rich in base saturation.
- These have high water holding capacity, although high (350-410 mm/m), yet a large part of it is not available for plant growth because held tenaciously by the dominant smectite clay. The workability of the soil is often limited to very short period of optimal water status.
- Being calcareous, they have pH ranging from 7.8 to 8.7 that may go up to 9.5 under sodic conditions. The $CaCO_3$ content increases irregularly with depth because of churning process; in some cases, it may form a calcic horizon within 1 m of soil surface.
- The soils do not exhibit any eluviation and/or illuviation process because of churning. Moreover, under calcareous conditions, the clay stays flocculated and doesn't move with percolating water.
- These have low bulk density because of the high clay content.
- These are very dark in colour which may due to clay-humus complexes and/or presence of titaniferous magnetic mineral.
- The soils are extremely hard, when dry and highly sticky and plastic, when wet and thus are difficult to cultivate and manage.
- These soils are commonly deficient in organic C, N, P and some extent of S.
- The soil develops shrinkage cracks, as the water from upper soil layers is depleted through evapotranspiration. The surface area of cracks may be 3 to 4.6 times the area of field. The water loss through evaporation from cracks may be as high as 55% of that evaporated at the surface. The developments of cracks also damage the roots.

Crop productive technology: Promising techniques which address the problem are now available. The same are discussed below

(A) Land configuration
1. **Broad bed and furrow system:** The system of grading of 90 to 150 wide beds across the contour to a 0.6 % slope. The beds are separated by furrow that drains into grassed water ways. The furrow about 50 cm wide and about 15 cm deep provide effective means of surface drainage to prevent water logging of all crops growing on the raised beds. The system reduces runoff and soil erosion under both fallow and crop condition.
2. **Graded furrow:** In black soil regions with 500 to 1000 mm rainfall, the productivity of upland rainy season crops can be substantially improved by providing furrows graded to 0.2 to 0.3 % slope to transmit excess rain water.
3. **Ridge and furrow system:** Planting of *kharif* crops on ridge about 15-20 cm high and suitably spaced on less than 1.2 % grade have been observed to improve crop stand, crop yields by as much as 27 to 106 % under high rainfall conditions (1280 mm).
4. **Raised and sunken bed system:** The system consists of all array of alternating raised and sunken beds. The runoff from raised beds planted to all upland crops is collected in the adjacent sunken beds, supporting rice or any other water tolerant crop. Under high rainfall areas and under low soil permeability conditions 3 m wide and 20 cm high or 6 m wide and 30 cm high beds with equally wide sunken bed, have been observed to yield the higher returns. The system ensures not only quick removal of excess rainfall but also favour recharge of the sub soil with rain water thus the system also alleviates the adverse effect of prolong dry spell on the crops.

(B) Mulching: Vertisols and soil with Vertic characters develops wide and deep cracks in dry seasons. The shrinkage cracks effect on root and water loss through evaporation may be as high as 55% of that evaporated at the surface. Mulching is useful as it reduce the development of cracks by checking water loss through evaporation.
1. Shallow cultivation with a blade harrow between a crop rows. It covers the cracks and act as effective mulch.
2. Tillage of surface soil (3.5 cm) and 1.5-2 cm thick wheat straw mulch on wet black soil has been observed to reduce evaporation by 30 %.
3. Sorghum stubble mulch @ 7.5 t/ha reduces the cracks and evaporation.
4. Use plastic mulch ten to two weeks preceding planting.

10.6. SANDY SOILS

Sandy soils have low content of clay and silt fractions and large preponderance of sand in their mechanical composition. The upper limit of USDA loamy sand textural class defined as the sum of silt per cent and one and half times the clay per cent not exceeding thirty seems to a good approximation of the allowable range. However, it is to be conceded that some marginal sandy loams may still be found problematic. Whereas those loamy sands dominated by very fine sand fractions may, in fact, be quite productive, without an extraordinary management. The soils qualifying as sandy in the region of our consideration usually contain less than 12% of silt and clay

Extent: The sandy soils are dominant in and as an associate feature in 5.5 M ha. In latter situation the sandy soils occurs in association with Sierozems or Calciorthids / Camborthids in the north and non-calcic brown or Ustochrepts in the east. Allowing for the presence of non-sandy soils, the net area of sandy soils is around 20 M ha. By far the major occurrence of sandy soils is in the arid zones.

Mineralogy: A number of investigators have shown the preponderance of quartz and muscovite in light fraction of fine sands in the soils from fluvio-aeolian plains. Feldspars, both orthoclase and plagioclase are present in small amounts only. In comparison, the sands from Rajasthan were found to contain quartz, orthoclase and plagioclase and hardly any muscovite. The heavy fraction which constitutes 2 to 4 per cent in Haryana and Punjab soils is made up of hornblende, biotite, chlorite, tourmaline, zircon and garnet. The clay fraction is made up mainly of illite and a variable proportion of smectite, Kaolinite, chlorite and attapulgite.

Numerous analyses show that pH of the soils ranges between 7.8 and 8.5. The CEC values, consistent with the low clay content are 2.5 to 7.0 $cmol(p^+)$/kg with calcium and magnesium as the dominant cation.

Constraints

1. **Low water retention capacity:** Arising from the large preponderance of macropores, the sandy soils are able to retain only small amount of water. Water retained by sandy soils is only half to one-third of that retained by the non sandy soils. This property makes the sandy soils droughty and prone to large deep percolation losses of water received during occasional heavy rainfall or irrigation.

Table 10.8: Water retention (e/w) at various tensions and available water content in soils.

Textural class	10 kPa	33 kPa	1.5 MPa	Available water*
Sand	9.1	5.3	1.6	7.5
Loamy sand	15.3	9.9	2.9	12.4
Sandy loam	18.5	11.3	4.7	13.8

* Available water = water content at 10 kPa – water content at 1.5 MPa

2. **Low unsaturated state hydraulic conductivity:** Understandably, the sandy soils posses high infiltration rate (5 to 20 cm/h) as well as high saturated hydraulic conductivity. These properties permit a rapid soaking in of rain and negligible runoff though it does create a problem of uneven distribution of water under irrigated management. However, under saturated condition, the behaviour of sandy soils is just reverse. The sandy soil is 200 times less conductive at 500 kPa tension than a sandy loam soil. The same is true for water diffusivity, which is ratio of hydraulic conductivity to the water capacity of the soil.

The low unsaturated hydraulic conductivity in these coarse textured soils has significant implications. Highly restricted water movement permits full utilization of the stored available moisture only if the plant roots are able to reach the sites where it is actually held. This is one reason why crops like cotton with pronounced tap root are not able to perform as well as those with fibrous root system. Further, since nutrient uptake by plant though mass flow or diffusion is with water as a medium, the low water conductivity restricts the utilization of native and applied nutrients.

Poor structure and high erodibility: The structure of soil is loose to very weakly developed depending upon clay content as can be seen from the following, based on field observation from a large number of profiles

Clay (%)	Structure
<5	Structure less, single grain
5-8	Very weak sub-angular blocky. Distinct peds are hardly observable. Weak clods do exist which crush almost entirely in to unaggregated mass.
9-12	Weak, sub-angular blocky. Some peds observable. Clods can stand some rough handling. Upon crushing only 10 to 25 per cent of the mass is in the form of peds.

Even a weak structure is quite significant as it endows the soil with a resistance to wind erosion. Since most sandy soils as seen from above are devoid of any structure development, these suffer from intensive erosion under prevailing wind regime of the desert region.

3. **Mechanical resistance:** Lack of cohesiveness enables the sandy soils to be worked at almost any water content. However, this also makes these soils prone to high compaction and development of plough pan where heavy machinery is used in farm operations.

4. **Low nutrient reserves:** Not only available but even the potentially available forms of nutrient elements are concentrated in clay, silt and organic fractions in soils. Numerous studies suggest a strong correlation between total as well as potentially available forms of nutrients with clay and/or organic matter. Since all the above fractions are conspicuously low in arid sandy soils, their fertility behaviour becomes understandable.

5. **Low nutrient diffusivity and buffering capacity:** Diffusion is the main transport mechanism for P, K and micronutrient uptake. The self diffusion coefficient of loamy sand soil is nearly one fourth of that in a loam soil. A high rate of nutrient application is necessary in these soils for adequate supply through diffusion. Considering diffusivity and buffering capacity, sandy soil require higher concentration of nutrient in solution for adequate plant nutrition than other soils. Though because of low adsorption, the sandy soils are able to attain this concentration with lower rate of fertilizer application, a situation is created for luxury consumption of nutrient in early stages of crop growth followed by tail-end inadequate supply and a depression to crop growth.

6. **Leaching losses of applied nutrients:** In whichever form N is applied in soil, it is converted to non-reacting nitrate form. Numerous studies indicated that the peak concentration of such solute occurs at a depth, the storage capacity of which is equal to depth of applied water. Thus far deeper movement and peak concentration in sandy soils are obvious. Even with a reactive nutrient like P, Sharma et al. (1982) noted the movement of surface applied P to a depth of 38 cm in a sandy loam soil with an application of 10 cm water.

Crop production technology in Sandy soils

(A) Increase clay/finer soil particles.
1. Application of pond sediments.
2. Use bentonite clay @ 2.5-3 kg/pit per metre of trench.
3. Mix fine textured soil.

4. Place asphalt barrier at shallow depth: These practices were found effective in reducing percolation losses, enhancing soil water storage and increased the crop yield by 20-40%.

(B) Soil compaction: Soil compaction with 200 kg roller by making 20 passing at proper moisture or 1500 kg roller, 4 passing.
   – Increase BD of 10-20 cm layer from 1.56 to 1.67 g/cc
   – Decrease infiltration rate by 25%
   – Reduce amount of irrigation water, each irrigation by 40%
   – Increase seedling emergence

(C) Mulches: Mulches are effective means to conserve soil water and moderate surface soil temperature. Improved physical environment due to mulches leads to better utilization of native and applied nutrients in surface soil greater seeding emergence, reduced weed growth and ultimate higher yields and water was efficiency.Organic mulches reduces the soil temperature as well as evaporation loss of water There are three types of mulches

1) Soil mulches

2) Straw mulch

3) Plastic mulch

Another effective treatment is solarization. It is a technique that involves trapping of solar energy by covering the soil with transparent polythene sheet during hot summer to increase soil temp. Primarily it has been used for control of soil bear pathogens', but also reduces weeds and improves mineralization.

(D) Methods of irrigation: With low stream sizes as available from wells and tube wells, the farmers in the area of sandy soils follow a check- basin system. Lining of channels with a paste of FYN and pond silt is done to reduce conveyance lasses. As an improvement sprinkles system of irrigation has been tried and the method has gained widespread papnlurily assist serves irrigation water depending upon type of crop and soil: Drip irrigation is also a good. Management practice to save water in arid and semi arid areas with limited and inferior quality of waters.

(E) Management of Applied Nutrients: The first irrigation of 5.5 and 9.5 cm leached, respectively, 50 and 81 per cent of nitrate nitrogen from top loan soil and 29 and 52 per cent from 105 cm depth. The trend was maintained with increasing number of irrigation.

Reduced leaching with light irrigation was marked by increased nitrogen uptake and significantly higher yields. Three spilts of fertilizer maintained higher no$_3$ status in top 60 cm soil and produced 30, 14 and 9 per cent higher yield of wheat then that from 1, 2 or 4 splits.

(F) Deep Tillage: Though sandy soils are porous and friable, poor structure makes them pronto formation of dense layer or plough sole which can hinder proliferation and penetration of root system. Deep tillage caused some reduction in bulk density (1.6 to 1.51 Mg m)

(G) Use of Amendments: Amendment of soil by the addition of fine textured materials is an age-old practice on irrigated holdings in the region. Application of pond sediments @ 152 to ha^{-1} had increased field capacity from 8.6 (control) to 11.5 per cent. Yield of pearl millet from 10.3 to 17.5 q ha^{-1} and reduced infiltration rate from 15.0 to 10.8 cm h^{-1} in a normal rainfall year

CHAPTER 11

FERTILIZER STORAGE AND FERTILIZER CONTROL ORDER

Warehousing and storage of fertilizers is a very important and massive activity. Ideally a marketer would like the fertilizer to spend minimum time in a godown because storage costs money, blocks money, occupies space, needs supervision and inspite of precautions, some fertilizer can be stolen or damaged. Storage can be called a necessary evil.People who pay for storage, often think whether it is better to spend on this item or to give off- season rebate to the farmer and let him do the storage.

11.1. THE PRINCIPLES OF GOOD STORAGE AT THE FIELD LEVEL ARE

(i) The fertilizers should be stored in a cool, dry and damp proof godown. The rain water must not get entered in the godwon and there is no need to have windows in the godown. But they should have proper ventilation for regulating for exit of gases from the store. The ventilators should be sealed in rainy season.

(ii) The bags should not be piled up directly on the floor as moisture of the floor causes the damage to the fertilizer. The wooden racks should be used for pilling the fertilizer bags.

(iii) The bags should not be piled together in a row of 8-10 bags.

(iv) The bags should not touch the wall of the godown.

(v) Proper space should be allowed between two of piled fertilizers for convenience of lifting the fertilizers.

(vi) The fertilizer that are hygroscopic in nature such as Urea, Ammonium Nitrate, Ammonium Sulphate Nitrate, Calcium Ammonium Nitrate must be stored in water proof bag and the entire bag should be used in one lot. Otherwise, the bag should be tied tightly and stored in a dry and damp proof godwon after taking required fertilizers.

(vii) The fertilizers that are fire hazardous such as Ammonium Sulphate must be handled very carefully.

(viii) All types of fertilizers such as Nitrogenous, Phosphatic and Potassic fertilizers should not be piled together. But they should be piled separately so that their handling is easy and gas fumes release from one group may not affect the quality of others.

(ix) The bag should not be kept open at any time to avoid the formation of cakes or lumps.

(x) The home mixed fertilizer should not be stored. Rather it should be used immediately after mixing of different fertilizers.

(xi) Prolonged storage of fertilizer should be avoided.

11.2. FERTILIZER CONTROL ORDER

The history of the Indian fertilizer industry dates back to 1906, when the first fertilizer factory opened at Ranipet (Tamil Nadu). Since then, there have been major developments in terms of both the quantity and the types of fertilizers produced, the technologies used and the feedstocks employed. The fertilizer industry in India is in the core sector and second to steel in terms of investment.

Prior to 1960/61, India produced only straight nitrogenous fertilizers [ammonium sulphate (AS), urea, calcium ammonium nitrate (CAN), ammonium chloride and single superphosphate (SSP)]. The production of NP complex fertilizers commenced in 1960/61. Currently, India produces a large number of grades of NP / NPK complex fertilizer. These include 16–20–20, 20–20–0, 28–28–0, 15–15–15, 17–17–17, 19–19–19, 10–26–26, 12–32–16, 14–28–14, 14–35–14 and 19–19–19. In addition, India produces various grades of simple and granulated mixtures.

The fertilizer was declared as an Essential Commodity in 1957 in India. To control the trade, price, quality of fertilizers and their distribution, "The fertilizers (Control) Order" came in to force in 1957. Since then the The Fertilizer (Control) order (FCO) has been amended periodically. It is useful for the personnels engaged in: Fertilizer manufacture, fertilizer business, fertilizer analysis and fertilizer inspection.

11.3. FERTILIZER LEGISLATION

Chemical fertilizers are becoming increasingly expensive day by day due to hike of prices of petroleum, inflation etc., which tempts dealers to adopt malpractices for earning more profits through adulteration, supplies of underweight materials or blending of degraded fertilizers etc. Thus, the farmers are ditched and often they fail to get good response of applied fertilizers. Therefore, the laws regulating the manufacture and sale of various fertilizers are essential to ensure that the consumer or the farmer is supplied with fertilizers of standard quality.

Keeping these points in mind, the Government of India brought in the fertilizer Control Act.

11.4. FERTILIZER CONTROL ACT

The Union Government of India promulgated the fertilizer Control Act (F.C.O) in 1957 under the Essential Commodities Act, 1955 (section 3) with a view to regulate fertilizer business in India.

The F.C.O. keeps a strict watch on quality control of fertilizers, provides for the registration of dealers and statutory control of fertilizer prices by Government. Therefore, everybody involved in fertilizer business as a manufacturer, dealer or a salesperson, must have proper understanding of the F.C.O. in order to avoid infringement of Government regulations.

The provisions given in the Order will also help the consumers/ farmers to know their rights and privileges in respect of fertilizer quality and Authorities to be approached for their grievances regarding supply of substandard materials, overcharging or containers of underweight supplies.

The F.C.O. is published by the Fertilizer Association of India (F.A.I.), updated when ever felt necessary. The Order has provisions on quality for each consumed fertilizer product and F.C.O. should be consulted under infringement of any of them.

Control of Quality of Fertilizers

The F.C.O. has provisions to penalize manufactures, distributors, and dealers for supply of spurious or adulterated fertilizers to consumers or farmers. The F.C.O. has fixed specifications for various fertilizers, which must be present in them failing which the legislation comes in force, and guilty is punished.

11.5. SPECIFICATIONS OF FERTILIZERS

To control the quality of fertilizers "The Fertilizer Control Order, 1985" has laid down specifications for the fertilizers. The parameters of the specifications are as follows:

i. Moisture, per cent by weight maximum
ii. Total nutrient content, percent by weight
iii. Forms of nutrient, per cent by weight
iv. Impurities, per cent by weight
v. Particle size.

Table 11.

1. Ammonium Sulphate

(i) Moisture per cent by weight, maximum	1.0
(ii) Ammoniacal nitrogen per cent by weight, minimum	20.6
(iii) Free acidity (as H2SO4.) per cent by weight, maximum (0.04 for material obtained from by product ammonia and by-product gypsum)	0.025
(iv) Arsenic as (As2O3) per cent by weight, maximum	0.01
(v) Sulphur (as S) ,per cent by weight, minimum	23.0

2. Urea (46% N) (While free flowing)

(i) Moisture per cent by weight, maximum	1.0
(ii) Total nitrogen, per cent by weight, (on dry basis) minimum	46.00
(iii) Biuret per cent by weight, maximum	1.5
(iv) Particle size—Not less than 90 per cent of the material shall pass through 2.8 mm IS sieve and not less than 80 per cent by weight shall be retained on 1 mm IS sieve	

3. Potassium Chloride (Muriate of Potash)

(i) Moisture per cent by weight, maximum	0.5
(ii) Water soluble potash content (as K_2O) per cent by weight, minimum	60.0
(iii) Sodium as NaCl per cent by weight (on dry basis) maximum	3.5
(iv) Particle size —minimum 65 cent of the material shall pass through 1.7 mm IS sieve and be retained on 0.25 mm IS sieve.	

4. Diammonium Phosphate (18-46-0)

(i)	Moisture per cent by weight, maximum	1.5
(ii)	Total nitrogen per cent by weight, minimum	18.0
(iii)	Ammonical nitrogen form per cent by weight, minimum	15.5
(iv)	Total nitrogen in the form of urea per cent by weight, maximum	2.5
(v)	Neutral ammonium citrate soluble phosphates (as P_2O_5) per cent by weight, minimum	46.0
(vi)	Water soluble phosphates (as P_2O_5) per cent by weight, minimum	41.0
(vii)	Particle size — not less than 90 per cent of the material shall pass through 4 mm IS sieve and be retained on 1 mm IS sieve. Not more than 5 per cent shall bebelow than 1 mm size.	

5. Zinc Sulphate Heptahydrate ($ZnSO_4.7H_2O$)

(i)	Matter insoluble in water per cent. by weight, maximum	1.0
(ii)	Zinc (as Zn) per cent. by weight, minimum	21.0
(iii)	Lead (as Pb) per cent by weight, maximum	0.003
(iv)	Copper (as Cu) per cent by weight, maximum	0.1
(v)	Magnesium (as Mg) per cent by weight, maximum	0.5
(vi)	pH not less than	4.0
(vii)	Sulphur (as S),percent by weight, minimum	10.0
(vii)	Cadmium (as Cd), percent by weight, maximum	0.0025
(ix)	Arsenic (as As),percent by weight, maximum	0.01

11.6. SPECIFICATIONS OF MANURE

Example : **Vermicompost :**

(i)	Moisture, per cent by weight	15.0-25.0
(ii)	Colour	Dark brown to black
(iii)	Odour	Absence of foul odour
(iv)	Particle size Minimum material should pass through 4.0 mm IS sieve	90%
(v)	Bulk density (g/cm^3)	0.7-0.9
(vi)	Total organic carbon, per cent by weight, minimum	18.0
(vii)	Total Nitrogen (as N), per cent by weight, minimum	1.0
(viii)	Total Phosphates (as P_2O_5), per cent by weight, minimum	0.8
(ix)	Total Potash (as K_2O), per cent by weight, minimum	0.8

(x)	C:N ratio	<20
(xi)	pH	6.5-7.5
(xii)	Pathogens	Nil
(xiii)	Conductivity (as dsm^{-1}),not more than	4.0
(xiv)	Heavy metal content, (as mg/kg), maximum	
	Cadmium (as Cd)	5.0
	Chromium (as Cr)	50.00
	Nickel (as Ni)	50.00
	Lead (as Pb)	100.00

Fertilizer Movement Control Order

The Fertilizer Movement Order (F.M.O.) was promulgated by Government of India in April 1973 to ensure an equitable distribution of fertilizers in various States. According to the fertilizer movement order, no person or agency can export chemical fertilizers from any State. However, Food Corporation of India, Warehousing Corporation of India and Indian Potash Limited; materials like Rock phosphate, bone meal (both raw and steamed) and zinc sulphate are exempted from the movement restriction.

Agency responsible for Enforcement of F.C.O

The Controller of Fertilizers for India, usually a Joint Secretary to the Government of India (Ministry of Agriculture) is responsible for the enforcement of F.C.O. throughout the country.

CHAPTER 12

FERTILIZER RECOMMENDATIONS AND APPLICATION

12.1. BLANKET RECOMMENDATION

Based on the fertilizer experiments conducted in different regions with improved varieties, fertilizer dose is recommended for each environment.

This approach does not consider soil contribution. However, it is suitable for recommendation of nitrogen since residual effect of fertilizer N applied to previouscrop is negligible and soils are generally low in nitrogen content.

Problem: Let the recommended fertilizer dose for low land rice be, 120, 60, 40kg N-P_2O_5 and K_2O per hectare, respectively. The amount of fertilizer required in the form of urea, single super phosphate (SSP) and muriate of potash (MOP) is calculated as shown below:

Urea contain 46%N

To supply 46kg N, 100kg urea is necessary

To supply 120kg N/ha, $\frac{100}{46} \times 120 = 260.9$ kg or 261 kg urea is required

Similarly,

SSP contain 16% P_2O_5

To supply 60kg P_2O_5/ha, $\frac{100}{16} \times 60 = 375$kg SSP is required

SOIL FERTILITY AND NUTRIENT MANAGEMENT

MOP contain 58% K_2O

To supply 40kg K_2O/ha, $\dfrac{100}{58} \times 40 = 68.9$ or 69kg MOP is required

Problem: In above example, fertilizer dose of paddy is 120, 60, 40kg N-P_2O_5 and K_2O per hectare, respectively. The recommendation of fertilizer is given below

Nutrient application

Category	N	P_2O_5	K_2O
Low	150	75	50
Medium	120	60	40
High	90	45	30

Fertilizer application

Category	Urea	SSP	MOP
Low	326	469	86
Medium	261	375	69
High	196	281	52

12.2. SOIL TEST CROP RESPONSE (STCR) APPROACH

In this approach, soilcontribution and yieldlevel are considered forrecommending fertilizer dose. This approach is also called as rationalized fertilizer prescription. From the soil test crop response experiments, following parameters are available.

Nutrient requirement(kg nutrient/q of grain) : $\dfrac{\text{Total uptake of nutrient (kg/ha)}}{\text{Grain yield (q/ha)}}$

% contribution from soil (CS) :

$$\dfrac{\text{Total uptake of nutrient in control plot(kg/ha)}}{\text{Soil test value of nutrientIn control plot (kg/ha)}} \times 100$$

Contribution from fertilizer (CF) $\dfrac{\text{Total uptake of nutrient in Treated plot}}{} - \dfrac{\text{Soil test value of nutrient In treated plot (kg/ha)}}{} \times \dfrac{CS}{100}$

% Contribution from fertilizer : $\dfrac{CF\ (kg/ha)}{Fertilizer\ dose} \times 100$

Fertilizer dose(kg/ha) :

$$\dfrac{\text{Nutrient requirement in kg/q of grain} \times 100}{\text{\% Contribution from fertilizer}} \times T - \dfrac{\text{\% contribution from soil}}{\text{\% contribution from fertilizer}} \times STV\ (kg/ha)$$

Based on this, fertilizer recommendations are developed for different regions. One such equation developed to recommend P and K, fertilizers for sugarcane in south Gujarat is given below:

Dose of P_2O_5 (kg/ha) = 2.24T - 3.97 × STV for available P_2O_5

Dose of K_2O (kg/ha) = 2.67T - 0.383 × STV for available K_2O

12.3. NUTRIENT USE EFFICIENCY (NUE)

"Nutrient use efficiency defined as yield(biomass) per unit input (Fertilizer, nutrient content)". The nutrient most limiting plant growth are N, P,K and S. NUE depends on the ability to efficiently take up the nutrient from the soil, but also on transport, storage, mobilization, usage within the plant and even on the environment. Two major approaches may be taken to understand NUE. Firstly, the response of plants to nutrient deficiency stress can be explored to identify processes affected by such stress and those that may serve to sustain growth at low nutrients input. A second approach makes use of natural or induced genetic variation.

Increasing nutrient efficiency is the key to the management of soil fertility. The proportion of the added fertilizer actually used by plants is a measure of fertilizer efficiency. Soil characteristics, crop characteristics and fertilizer management techniques are the major factors that determine fertilizer efficiency.

Factors influencing nutrient use efficiency (NUE)

Soil characteristics

(1) **Nutrient Status of Soil:** The response of any crop or a cropping system to added nutrient depends largely upon the inherent capacity of soil to supply that nutrient as per the requirement of crop. In a low nutrient soil, the crop responds remarkably to its application. On the other hand, in a high nutrient

soil, the crops may show little or no response. In medium test soil, the response is intermediate. Soil testing helps in adjusting the amount of fertilizer and thus improves the efficiency of fertilizers use. By demarcating the areas responding differently to different plant nutrients, right type and proper amount of fertilizers can be applied to them.

(2) **Nutrient Losses and Transformations:** The amounts of nutrients estimated by soil tests may not be entirely available to plants because of their leaching, volatilization, denitrification and transformations to unavailable forms. Leaching losses are important for nitrate nitrogen because it is not held by exchange sites in the soil, it is lost. Such losses are of particular significance in sandy soils and in situations if heavy rain or irrigation follows its application. In acid soils, leaching losses of calcium, sulphate, potassium and magnesium are more common. Volatilization of ammonia in high pH surface soils is considerable when urea is applied at the surface. Denitrification loss of nitrogen mainly occurs under waterlogged conditions prevailing during rice cultivation, particularly under higher temperatures and in the presence of easily decomposable organic materials.

The conversion of a portion of available nutrients into insoluble mineral forms is also important. Thus, the efficiency of added phosphorus is 20 to 30 per cent. Microbial immobilization also converts temporarily the soluble forms of nutrients into unavailable forms. Similarly, the efficiency of zinc applied to soil is less than 3%.

Soil characteristics play a dominant role in the transformation of nutrients. Soil reaction (pH) is one of the important soil properties that affects plant growth. The harmful effects of soil acidity are more due to secondary effects except in extreme case. The important secondary effects of high acidity or low pH in a soil are the inadequate supply of available calcium, phosphorus and molybdenum on one hand and the excess of soluble aluminum, manganese and iron on the other. Likewise, in saline-alkali soil, the deficiency of Ca, Mg, P, Zn, Fe and Mn is very common. The fertilizers practices are, therefore, to be modified accordingly for soils with different soil reactions. The main aim of liming of acid soils and addition of gypsum to alkali soils is to change the soil pH suitable for the availability of most plant nutrients.

(3) **Soil Organic Matter:** Soil organic matter content is generally considered as the index of soil fertility and sustainability of agricultural systems. It improves the physical and biological properties of soil, protects soil surface from erosion and provides a reservoir of plant nutrients. In tropics, the maintenance of soil organic matter is very difficult because of its rapid decomposition under high temperatures. The cultivation of soils generally decreases its organic carbon content because of increased rate of decomposition by the current agricultural

practices. In cultivated soils, prevalent cropping system and associated cultural practices influence the level at which organic matter would stabilize in a particular agro-eco-system. Long-term fertilizer experiments have shown that the integrated use of organic manures and chemical fertilizers can maintain high productivity and sustainable crop production. Recent studies have indicated that a periodic addition of large quantity of crop residue to the soil maintains the nitrogen and organic matter at adequate levels even without using legumes in the rotation. The application of FYM, compost and cereal residues effectively maintains the soil organic matter. There is a significant increase in soil organic matter due to incorporation of rice or wheat straw into the soil instead of removing or burning it. Yields are, however, low in residue incorporated treatments due to wide C:N ratio of the residues. This ill effect, however, can be avoided if the rice straw is incorporated at least 20 days before seeding wheat.

(4) Soil moisture: Fertilizer application facilitates root extension into deeper layers and leads to grater root proliferation in the root zone. Irrigated wheat fertilized with nitrogen used 20-38 mm more water than the unfertilized crop on loamy sand and sandy loam soils and increased dry matter production Soil moisture also affects root growth and plant nutrient absorption. The nutrient absorption is affected directly by soil moisture and indirectly by the effect of water on metabolic activities of plant, soil aeration and concentration of soil solution. If soil moisture becomes a limiting factor during critical stage of crop growth, fertilizer application may adversely affect the yield.

(5) Physical Conditions of Soil: Despite adequate nutrient supply, unfavorable physical conditions resulting form a combination of the size, shape, arrangement and mineral composition of the soil particles, may lead to poor crop growth and activity of microorganisms. Soil nitrogen generally increases as the texture becomes finer. The basic requirements for crop comes finer, The basic requirements for crop growth in terms of physical conditions of soil are adequate soil moisture and aeration, optimum soil temperature and freedom from mechanical stress. Tillage, mulching, irrigation, incorporation of organic matter and other amendments like liming of acid soils and addition of gypsum to sodic soils are the major field management techniques that aim at creating soil physical environment suitable for crop growth. Tillage affects water use by crops not only through its effect on root growth but also affects the hydrological properties of soils. Mulching with residues, plastic film *etc.,* influences evaporation losses from soil by modifying the hydro-thermal regime of the soil and affects root growth and rooting pattern. Use of organic mulch also decreases maximum soil temperature in summer and increases minimum soil temperature in winter and help in the conservation of soil moisture.

Crop Characteristics

(i) Nutrient Uptake: The total amount of nutrients removed by a crop may not serve as an accurate guide for fertilizers recommendations; it does indicate the differences in their requirement among crops and the rate at which the nutrients reserves in the soil are being depleted. The nutrient uptake may vary depending upon the crops and its cultivars, nutrient level in the soil, soil type soil and climatic conditions, plant population and management practices. It is estimated that 8t of rice grain remove 160 kg N, 38 kg P, 224 kg K, 24 kg S and 320 g Zn as compared to a removal of 125 kg N, 20 kg P, 125 kg K, 23 kg S and 280 g Zn by 5t of wheat from one hectare field.

(ii) Root Characteristics: Roots are the principal organs of nutrient absorption. A proper understanding of their characteristics helps in developing efficient fertilizer practices. The absorption of nutrients depends upon the distribution of roots in soil. The shallower the root system, the more dependent the plant is on fertilizers. Hence, any soil manipulation, which encourages deep rooting, will encourage better utilization of fertilizers. It is well known that some plants are better scavengers of certain nutrients than others. This is mainly because of the preferential absorption of these nutrients by the roots of those plants. For example, legumes have a marked preference for divalent cations like Ca^{2+} whereas grasses feed better on monovalent cations like K^+.

The efficiency of the applied fertilizer can be improved considerably if the rooting habits of various plants during early growth stages are known. This is particularly true for relatively immobile nutrients and for situations where the fixation of applied nutrients is very high. If a plant produces tap root system early, fertilizer can best be placed directly below the seed. On the other hand, if lateral roots are formed early, side placement of fertilizer would be helpful.

Mycorrhizal fungi often associated with plant roots, increase the ability of plants to absorb nutrients particularly under low soil fertility. However, fertilizer additions generally reduce their presence and activity.

Crop Rotation: The nature of cropping sequence has a profound effect on the fertilizer requirement and its efficiency. Crops are known to differ in their feeding capacities on applied as well as native nutrients. The crops requiring high levels of fertilizers such as maize, potato may not use the applied fertilizers fully and some amount of the nutrient may be left in the soil which can be utilized by the succeeding crop. Phosphorus, among the major nutrients, is worthy of consideration because only less than 20 per cent of the applied phosphatic fertilizer is utilized by the first crop. Similarly, less than 3% of the applied zinc is used by the first crop. The magnitude of the residual effect is, however, dependent on the rate and kind of fertilizer used, the cropping and management system followed and to a great extent on the type of soil. Crops have a tendency of luxury consumption of N and K and may not leave any residual effect unless doses in excess of the crop requirement are applied. On the other hand, if sub-optimal doses of fertilizers are applied to a crop, they may leave the soil in a much exhausted condition and the fertilizer requirement of the succeeding crop may increase. The legumes leave nitrogen rich root residues in the soil for the succeeding crop and thus reduce its nitrogen requirement.

12.4. FERTILIZERS MANAGEMENT UNDER RAINFED CONDITIONS

In dry land agriculture, limited water availability is usually the factor that ultimately limits crop production. However, it is not unusual for limited availability of one or more soil nutrients to further decrease production potential. Often, the effects of water and nutrient deficiencies are additive. Because soil used under dry land agriculture is developed under widely varying conditions, their ability to supply nutrients is highly variable.

Fertilizer practices greatly affect nutrient cycling and availability in rainfed conditions. Because of frequent dry periods, placement of soluble fertilizers with the seed is extremely hazardous in dryland soils. The higher rates of fertilizer application may result in high osmotic potentials near the germinating seed. For oil crops, applying no fertilizer N with the seed is usually recommended. However, up to 20 to 30 kg P/ha can be applied with the seed because of the considerably lower solubility of most P fertilizer. It is also reported that P availability is particularly critical for an eroded soil.

In dryland soils, the surface layers often remain dry for a major part of the growing season. Such a condition might suggest that fertilizers should be placed deeper in the region of the active root zone for more of the growing season.

Timing of fertilizer application could also affect nutrient cycling. Applying N fertilizers near the time of maximum N uptake rate of the crop results in the most efficient uptake of the fertilizer.

Fertilizer sources also determine the growth the crops under rainfed conditions. Most dryland experiments showed that ammonium nitrate is usually one of the most efficient N sources for dryland crops. At the other extreme, these experiments showed that urea is the least efficient form of N fertilizers. One must exercise considerable caution when using urea on dryland to avoid excessive losses by ammonia volatilization.

By concentrating the urea (liquid or solid) in a band or pellets, surface contact is reduced, reducing volatilization. Injecting or incorporating urea beneath the soil surface is by far the best way in which to apply this material to dryland soils.

12.5. INTEGRATED PLANT NUTRIENT MANAGEMENT SYSTEM (IPNMS)

Definition

IPNM is the intelligent and combined use of inorganic, organic and biological resources so as to sustain optimum yields, improve or maintain the soil chemical and physical properties and provide crop nutrition packages which are technically sound, economically attractive, practically feasible and environmentally safe. The principal aim of the integrated approach is to utilize all the possible sources of plant nutrition in a judicious and efficient manner.

Concept of IPNMS

The basic concept of IPNMS is the promotion and maintenance of soil fertility for sustaining crop productivity through optimum use of all possible sources of nutrients like organic, inorganic and biological in an integrated manner appropriate to each farming situation. Improvement of soil fertility and productivity on sustainable basis through appropriate use of fertilizers and organic manures is the key principle and their scientific management for optimum growth and yield of crops in a specific agro ecological conditions.

Main objectives of IPNM or INM

1. To reduce the dependence on chemical fertilizers.
2. To maintain productivity on sustainable basis without affecting soil health.
3. To conserve locally available resources & utilize them judiciously.
4. To reduce the gap between nutrients used & nutrients harvested by the crop.
5. To improve physical, chemical & biological properties of soil.
6. To make soil healthy by providing balanced nutrients through different nutrient sources.
7. To overcome or reduce the ill effects of continuous use of only inorganic chemical fertilizers.
8. To improve economical status of farmers.
9. To increase the fertilizer use efficiency (FUE).

Components of INM

IPNM mainly emphasizes the integrated use of all the essential nutrients from different sources like chemical fertilizers, organic manures, green manures, bio-fertilizers, legume crops, locally available plant resources in a balanced proportion for sustainable soil health and productivity.

I. **Use of inorganic fertilizers:** They are very important for sustaining and increasing food production. Different kinds of fertilizers are commercially available in the market for all the major and micronutrients. However, they are costly inputs and their excessive use may deteriorate the soil quality and food quality. Hence, there is a need to improve their use efficiency through efficient and balanced fertilizer management and essentially follow the **four R's** formula for judicious and effective nutrient/fertilizer management. They are

Right Type of fertilizers.

Right Dose of fertilizers.

Right Method of application.

Right Time of application.

II. **Use of organic manures/ materials:** Due to intensive cultivation of soil and less organic manure application, the soils are low in organic matter status. A decrease in soil organic matter results in compact soil, poor aeration and low infiltration and water holding capacity and also low fertility status. The organic matter status in soils can be improved and maintained by constant addition of organic manures such as FYM. compost, green manures, poultry manures, vermicompost, oilcakes etc., Organic matter is good source of macro and micro nutrients, and more over improves physical, chemical and biological properties soil.

III. **Use of biological sources/biofertilizers:** Biofertilizers are cultures of micro organisms (bacteria, fungi, algae). Their use benefits the soil and plants growth by providing N & P and also brings about the rapid mineralization of organic materials in soils. They are capable fixing N, solubilizing and mobilizing the phosphorus and mineralizing organic matter in soil. Their incorporation improves the physical and biological properties of soils.

IV. **Maintaining the physical properties of soil:** Physical properties such as soil aggregation, soil texture, structure, aeration, water holding capacity (WHC), infiltration rate, etc., should be maintained regularly through better cultivation practices and organic manure applications to maintain soil fertility & nutrient availability.

V. Management of problematic soils: Problematic soils such as acid soils, saline and alkaline soils, water logged soils are known to decrease the productivity of the soil. Acid soil having the problems like toxicities of Iron, Mn, Al, deficiency of P & Mo. Similarly, saline and alkali soils showing the deficiency of Fe, Mn, Zn and Cu and also toxicities of Mo. These soils should be regularly managed and reclaimed through the application of soil amendments such as lime for acid soil, gypsum for alkali soils and other organic and inorganic materials based on soil test results. It helps to improve soil fertility and productivity and sustain the yield.

VI. Better/Judicious water management practices: Plants absorb the nutrients from the soil only in a dissolved state and sufficient moisture is therefore required for utilizing the nutrients of the soil. Management of moisture in the soil by improved and modern irrigation techniques like drip or sprinkler or basin where the rainfall is low and draining the soil where it is subjected to stagnation of water helps to increase water and nutrient availability to the crops.

12.6. SITE-SPECIFIC NUTRIENT MANAGEMENT

Recent advances in technology have combined the Global Positioning System (GPS) of earth-obtaining satellites, computer programs capable of making detailed maps that integrate information about many soil properties, and technologies that allow farm equipment to alter the rate of fertilizer, seed or chemical delivery on the go. The combined technology allows farmers to make nutrient management more site-specific than was previously practical for large farming operation. Portable GPS receivers can plot one's exact location (to within a few metes) as one moves across a large field. Therefore, if one is taking soil samples, the location of each sample can be geo-referenced with north-south and east-west coordinates.

In practice, a large field is divided into cell in a grid pattern, each cell usually being about 1 ha in area. For example, 18 separate geo-referenced soil samples (each being a composite of about 15 to 20 soil cores) may be collected from a single 18 ha field. These soil samples can be analyzed for each properties as texture, OM content, pH and soil test levels for P and K. A computer program can then produce a map of the spatial distribution of each soil property measured. Special statistical methods are used to estimate the location of continuous boundaries between areas differing in regards to each soil properties. Thus, one map might show area of low, medium and high soil test P levels. Another map might show areas of high, medium and low clay content and so on. Information on other spatial variables, such as soil classification, drainage class, past management, crop cultivars. A sophisticated computer program then integrates the information from these individual maps to crate a new combined map showing the different application rates of fertilizer (or other material) that are recommended for different part of the field.

For example, areas mapped as testing low as P might be stated to receive higher-than-average application rates of P fertilizer, while areas mapped as being high in P might receive no P fertilizers at all. Similarly, reduced rate of N fertilizer application might be mapped for areas of sandy soils with high leaching potential but low yield potential. The application rate recommendation maps area then programmed into a computer on board the machine that spreads the fertilizer. The fertilizer spreader has a GPS receiver installed, so its computer tracts its location on the map as it moves across the field. The computer then signals small motors in the fertilizer spreader to increase or decrease the rate of application, as called for on the map. Thus, the system is designed to reduced both over application on the high fertility areas (which should save money and reduce nutrient pollution potential) and under application on the low-fertility sites (which should make money by producing high yield). The total amount of fertilizer added may not be greatly different from that used in the past, but application rates should be more in turn with plant needs and environmental cautions.

At harvest time, similar computerized satellite linkages are used to monitor yields in different part of the field and create maps showing the yield differences. The yield maps produced by these GPS equipped combines are proving to be the most popular component of the entire system; as they help farmers pinpoint problem areas for investigation. Yield levels in some parts of a field, even in one that is uniform in appearance, commonly are 2 to 3 times as high as in other parts of the fields. By overlaying the yield map on the soil nutrient maps, it is possible to determine the extent to which nutrient decencies are constraining yields, but other problems such as poor drainage, low organic matter or even feeding by wildlife may also come to light.

This site specific nutrient management system is an integral part of what is commonly referred to as precisian farming. Opportunity for the control of insects and weeds and for modifying plant-seeding rates and depths can be utilized on site-specific basis rather than on field basis.

CHAPTER 13

ASSESSMENT OF IRRIGATION WATER QUALITY

All natural waters used for irrigation contains inorganic salts in solutions which are derived originally from the rocks or solid phase material through which water percolates. The most common dissolved constituents are chlorides, sulphates and bicarbonates of Ca, Mg and Na. The concentration and proportion of these salts determine the suitability of water for irrigation. Other constituent such as B, Li, F or other ions, which have a toxic effect on plants, may occur in lesser amounts in irrigation water. If water used for irrigation contains excessive quantities of the constituents noted above, it might affect the growth of plants in three ways viz.,

(a) As a result of adverse changes in the physical characteristics of the soil,
(b) The increased osmotic pressure of the soil solution may decrease the physiologicalavailability of moisture to plants,
(c) Accumulation of certain ions in the soil solution that may have a specific toxic effect upon the physiological processes of the plant.

Crop production in the arid and semi-arid regions is dependent on irrigated agriculture. The hot and dry climates of these regions require that the irrigation water does not contain soluble salts in amounts that are harmful to the plants or have an adverse effect on the soil properties. Water of such quality is usually not available in sufficient quantities to satisfy the water requirements of all the crops grown. Under these conditions the farmers are use irrigation water with high quantities of dissolved salts, invariably accompanied by yield reductions of most crops. use of such water can often lead to crop failures and to the development of saline or sodic soils which, in turn, require expensive treatment to make them productive again. On the other hand, when saline water is skillfully used, it can contribute to the successful production of a variety of crops.

The original source of salts in irrigation water is the rock that forms a part of the earth's crust - it is constantly subject to weathering which releases salts to be carried away by water. When the soil becomes truly saline, the visible surface evidence might be a white crust or dark, moist, oily looking patch. However, salt accumulation begins to affect crop yields long before visible signs of its presence appear.

There are three principal problems that can arise from the quality of irrigation water delivered to the agricultural fields.

13.1. QUALITY OF IRRIGATION WATER

1. **Salinity hazard:** This is directly related to the quantity of salts dissolved in the irrigation water. All irrigation water contains potentially injurious salts and nearly all the dissolved salts are left in the soil after the applied water is lost by evaporation from the soil or through transpiration by the plants. Unless the salts are leached from the root zone, sooner or later they will accumulate in quantities which will partially or entirely prevent growth of most crops.

Table 13.1: On the basis of EC, waters were divided into four classes by USSSL as follows:

Conductivity (micromhos/cm at 25 °C)	Class	Symbol	Inference
0 – 250	Low salinity	C_1	i) Can be used for most soil for most crops
			ii) Little likelihood of salinity
250 – 750	Medium salinity	C_2	i) Can be used with moderate leaching ii) Moderate salt tolerant crops should be grown
750 – 2250	High salinity	C_3	i) Cannot be used where drainage is restricted
			ii) Salt tolerant plant and additional management practices should be followed
2250 – 5000	Very high salinity	C_4	i) Not suitable for irrigation
			ii) Can be used occasionally with leaching
			iii) Salt tolerant crop should be grown with additional management practices

According to USSSL, the water having more than 5000 micromhos EC value cannot be used for purposes of irrigation.

2. **Alkali hazard:** This is another problem often confronting long-term use of certain water for irrigation and relates to the maintenance of adequate soil permeability so that the water can infiltrate and move freely through the soil. The problem develops when irrigation water contains relatively more sodium ions than divalent calcium and magnesium ions while the total concentration of salts is generally not very high. Accumulation of sodium ions on to the exchange complex results in a breakdown of soil aggregates responsible for good soil structure needed for free movement of water and air through the soils. As in the case of sodic soils, accumulation of sodium on the exchange complex can be reduced by applying appropriate quantities of amendments, e.g. gypsum.

The continuous use of water having high concentration of Na will convert a normal soil into an alkali soil. The sodium adsorption ratio (SAR) developed by USSSL expresses the relative activity of Na ions in cation exchange reactions with the soil. The exchangeable Na percentage (ESP), which the soil will attain when the soil and water are in equilibrium, can be predicted approximately from the value of SAR of water. Accordingly, the waters are divided into four classes with respect to the Na hazards as follows :

Table 13.2: Classification of Irrigation on the basis of SAR of water

SAR value	Class	Symbol	Inference
0 – 10	Low Na water	S_1	i) Can be used for all soils with little danger of harmful Na level development
			ii) The Na sensitive crops are affected
10 – 18	Medium Na water	S_2	i) Sodium hazard likely in fine textured soil
			ii) Can be used on soils having high permeability
18– 26	High Na water	S_3	i) May produce harmful level of exchangeable Na in most soils except gypsiferous soils
			ii) Requires special management practice like good drainage, high leaching and addition of organic matter and gypsum
More than 26	Very high Na water	S_4	i) Unsatisfactory for irrigation except at low and perhaps medium salinity of irrigation water, special management as above should be made

ASSESSMENT OF IRRIGATION WATER QUALITY

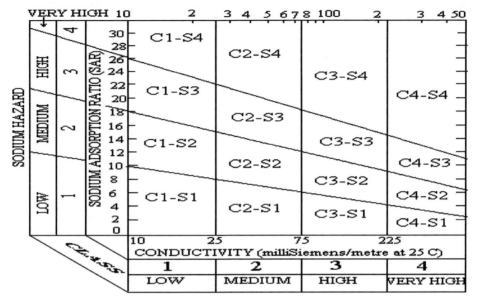

The USSSL has prepared the diagram for use of water having different values of EC as well as SAR.

3. **Bicarbonate hazard:** The bicarbonate ions are primarily important because their tendency to precipitate Ca and to some extent Mg, in the soil solution as their normal carbonates e.g. $Ca + 2HCO_3 \rightarrow CaCO_3 + CO_2 + H_2O$

The CO_3^{-2} ions are seldom present in water but HCO_3^{-1} ions may be present in appreciable proportion of the total anions present in irrigation waters. Based on the theory of precipitation of Ca and Mg, Eaton (1950) suggested the concept of "Residual Sodium Carbonate" commonly known as RSC. The RSC can be found out by following equation :

$$RSC = (CO_3^- + HCO_3^-) - (Ca^{++} + Mg^{++})$$

Where; concentrations of all ions are expressed in meq/lit.

It is obvious from above equation that as the Ca and Mg are lost from the soil solution by precipitation, the relative proportion of Na remaining in water is increased. Thus, the alkali hazard as defined by the SAR is increased. The standard for RSC as given by USSSL as follows :

Table 13.1:2 Classification of imagination center other basis of RSC (Meq/lit)

RSC (meq/lit)	Quality of irrigation water
Less than 1.25	Probably safe for most purpose
1.25 – 2.50	Marginal can be used on light textured soil with adequate leaching and application of gypsum
More than 2.50	Not suitable for irrigation purposes

4. **Boron hazard:** Boron is very toxic to plants at low concentration in the soil solution. Because boron tends to accumulate in the soil from even low concentration in the irrigation waters, it is necessary to consider this constituent in assessing the quality of irrigation waters. The USDA has suggested the type of crops to be grown with respect to boron content in irrigation water. The limits are as under :

Boron content of irrigation water (ppm)	Boron tolerance of crops	Crops to be grown
0.3 – 1.0	Sensitive	Citrus, Apricot, Peach, Apple, Pear, Plum, walnut
1.0 – 2.0	Semi-tolerant	Sweet potato, Oats, Sorghum, Maize, Wheat, Barley, Radish, Peas, Tomato, Cotton, Potato, Sunflower
2.0 – 4.0	Tolerant	Carrot, Cabbage, Onion, Beans, Sugarbeet, Alfalfa, Date

5. **Other hazards (Toxicity hazard):** A third problem results from the existence, in some water, of such toxic substances as boron or heavy metals. Boron, though an essential element for plant growth and nutrition, is required only in very small amounts. A high concentration of boron in the irrigation water can have a toxic effect on the growth of many plants. Similarly, certain other ions, e.g. chloride, sodium, etc., could prove toxic to specific crops if present in excessive quantities.

   (i) **Chlorides :** The grading of irrigation waters based on chloride content as proposed by Schofield is as under :

Chloride (meq/lit)	
0 – 4	Excellent
4 – 7	Good
7 – 12	Permissible
12- 20	Doubtful
More than 20	Unsafe

   (ii) **Fluorides :**

Fluoride (ppm)			
(i)	<1	Low	Safe
(ii)	1-2	Medium	Marginal
(iii)	>2	High	Unsafe

   (iii) **Other elements :** The safe limit for other elements present in irrigation water is as follows :

Element	For waters used continuously on all soil (ppm)	For used upto 20 years on fine textured soil at pH 6.0 to 8.5 (ppm)
Al	5.00	20.00
Arsenic	0.10	2.00
Cu	0.20	5.00
Fluorine	1.00	15.00
Lead	5.00	10.00
Lithium	2.50	2.60
Mn	0.20	10.00
Mo	0.01	0.05
Se	0.02	0.02
Zn	2.00	10.00
Fe	5.00	20.00

(iv) Pollutants : Many substances that are discharged as industrial wastes into surface streams may have phytotoxic properties. Great caution should be exercised in the use of irrigation water that is suspected of containing phytotoxic pollutants.

6. **Soluble sodium percentage (SSP):**

   $$\text{Soluble Sodium Percentage (SSP)} \frac{Na}{Ca + Mg + Na} \times 100$$

   Where; all soluble cations are expressed in me l^{-1}

   Irrigation waters having SSP value of 60 and above are considered as harmful.

7. **Magnesium hazard:** It is believed that one of the important qualitative criteria in judging the irrigation water is its magnesium content in relation to total divalent cations, since high magnesium adsorption by soils affects their physical properties. A harmful effect on soils appears when Ca:Mg ratio declines below 50.

   $$\text{Mg-adsorption ratio} = \frac{Mg^{2+}}{Ca^{2+} + Mg^{2+}}$$

   Magnesium hazard in irrigation water is expected having Mg:Ca ratio more than 1.

8. **Nitrate concentration:** Very frequently ground waters contain high amount of nitrate. When such type of irrigation water is applied on soils continuously various physical properties will be affected very badly which causes poor growth of the plants.

9. **Lithium:** Lithium is an important trace element which may be found in most of saline ground waters and irrigated soils. It has been found that small concentrations (0.05-0.1 ppm) of lithium in irrigation water produced toxic effects on the growth of citrus crops. It has also been reported that saline soils of varying degrees found in India contain lithium upto 2.5 ppm. Fortunately

the germination of majority of crops including rice, wheat, barley, etc. is not affected with this level of lithium content in soils. However, guidelines for the quality of irrigation water considering various criteria are presented in Table.

Table 13.4: Guidelines for irrigation water quality established by FAO

Water constituent	Intensity of problem*		
	No problem	Moderate	Severe
Salinity (dSm^{-1})	< 0.75	0.75-3.0	> 3.0
Permeability (rate of infiltration affected)			
Salinity (dSm^{-1})	> 0.5	0.5-0.2	< 0.2
Adjusted SAR; soils are			
Dominantly montmorillonite	< 6	6-9	> 9
Dominantly illite-vermiculite	< 8	8-16	> 16
Dominantly kaolinite-sesquioxides	< 16	16-24	> 24
Specific ion toxicity			
Sodium (as adjusted SAR)	< 3	3-9	> 9
Chloride (meq l^{-1})	< 4	4-10	> 10
Boron (meq l^{-1})	< 0.75	0.75-2.0	> 2.0
Miscellaneous			
NO_3^--N or NH_4^+-N (meq l^{-1})	< 5	5-30	> 3.0
HCO_3^- (meq l^{-1}) as damaged by overhead sprinkler	< 1.5	1.5-8.5	> 8.5
pH	6.5-8.4		0-5, 9.5+

*Based on the assumptions that the soils are sandy loam to clay loams, have good drainage, are in arid to semi-arid climates, irrigation is sprinkler or surface, root depths are normal for deep soil, and the guidelines are only approximate.

The CSSRI, Karnal has recommended following limits of EC for use of saline waters:

Table 13.5: Guidelines for interpretation of water quality for irrigation under Indian conditions:

Soil	Crops to be grown	Upper permissible limit of EC of water for safe use for irrigation, (dS/m)
Deep black s and alluvial.clay content > 30 percent. Soils that are fairly to moderately well drained.	Semi-tolerant	1.5
	Tolerant	2
Heavy textured soils clay content (20-30%) Soils that are well drained and have a good surface drainage system.	Semi-tolerant	2
	Tolerant	4
Medium textured soils . clay content (10-20%.) Soils that are very well drained and have a good surface drainage system.	Semi-tolerant	4
	Tolerant	6
Light textured soils having a clay content< 10%. Soils that have excellent internal and surface drainage.	Semi-tolerant	6
	Tolerant	8

Management practices for efficient use of high saline water: It would thus seem that there can be very wide variations in the permissible limits of salinity levels of water for irrigation. For this reason any rigid generalizations may prove disadvantageous for field level workers and there is need to develop guidelines for each major area having similar soil, climatic and agricultural conditions. More important however is our ability to use a water of a particular salinity level under a given set of conditions. Management practices can often be modified to obtain a more favorable distribution of salts in the profile and therefore better crop yields, water quality remaining the same. Management practices that can help to overcome a high salinity problem of the irrigation water are discussed below. Desalinization of water to remove soluble salts has often been referred to as a technical possibility but at the present stage of available technologies it is doubtful if this method can have any large-scale application in the utilization of saline water for irrigation of most agricultural crops, at least in the near future.

Management practices

1. **More frequent irrigation:** The adverse effects of the high salinity of irrigation water on the crops can be minimized by irrigating them frequently. More frequent irrigations maintain higher soil water contents in the upper parts of the root zone while reducing the concentration of soluble salts. Both these factors result in reduced effect of high salts on the availability of water to plants and therefore promote better crop growth. The sprinkler method of irrigation is generally more amenable to increased frequency of water applications. In surface irrigation methods however, more frequent irrigations almost invariably result in an appreciable increase in water use.

2. **Selection of salt tolerant crops and varieties:** As indicated in previous sections, there is a wide range in the relative tolerance of agricultural crops to soil salinity. Proper choice of crops can result in good returns even when using high salinity water, whereas use of such water for growing a relatively salt-sensitive crop may be questionable. Similarly, selection and breeding of salt-resistant crop varieties offer tremendous possibilities of utilizing saline water resources for crop production. Some workers have suggested induction of salt tolerance by soaking seeds for a certain period in salt solutions as a method for obtaining increased yields in saline water irrigated soils, while others suggest that growing seeds obtained from parents that have been irrigated with saline water helps in obtaining higher crop yields. These suggestions, however, have not been tested extensively on a field scale.

   Salt tolerance of different crops.

High tolerant : Rice, cotton, sugar beet, tobacco, date palms, ber.

Moderate tolerance : sugarcane, wheat, guava, pomegranate, tomato, sweet potato.

Highly sensitive : Beans, pea, grape, orange, apple, pear, carrot.

3. **Use of extra water for leaching:** To prevent excessive salt accumulation in the soil, it is necessary to remove salts periodically by application of water in excess of the consumptive use. The excess water applied will remove salts from the root zone provided the soil has adequate internal drainage. This concept (Richards, 1954) is quantified in the term 'leaching requirement' often referred to by the abbreviation, LR. By definition, leaching requirement (LR) is the fraction of total water applied that must drain below the root zone to restrict salinity to a specified level according to the level of tolerance of the crop.

$$LR = \frac{Ddw}{Diw}$$ where D is the depth of water,

and dw and iw refer respectively to the drainage and irrigation water. Assuming strict salt balance conditions in the soil-water system:

Diw x Ciw = Ddw x Cdw where C refers to the concentration of salts.

Therefore,

$$LR = \frac{Ciw}{Cdw} \text{ or } \frac{ECiw}{ECdw}$$

This would imply that the excess amount of irrigation water of a known EC that must be applied is determined by the maximum permissible EC of the drainage water specified for a particular crop. The values of ECdw represent the maximum salinity tolerated by the species grown under particular conditions. The leaching requirement for a particular crop may be illustrated by use of salt tolerance data (Figure 9). For barley, where a value of ECdw = 8 dS/m can be tolerated, leaching requirement = ECiw/8. Thus for irrigation water with conductivities of 1, 2 and 4 dS/m respectively, the leaching requirement will be 12, 25 and 50 percent.

In actual irrigation practice, the applicability of the leaching requirement concept has had some limitations. In the normal surface irrigation methods there are invariably 10 to 20 percent or more losses due to deep percolation of water beyond the root zone in most light and medium textured soils and this takes care of the leaching requirements for salinity control. In heavy textured soils and in soils having expanding type clay minerals applying 15 to 20 percent more water is often difficult during the crop season due to poor permeability and consequent aeration problems. Leaching accomplished periodically through seasonal rainfall may also result in adequate salt removal from the root zone.

Application of excess water, above that needed for meeting the evapotranspirational needs, though useful for salinity control, puts a high demand on the water resources on the one hand and increases the salt load of the drainage water on the other. It therefore appears that controlling the interval between irrigations is the most important management practice for obtaining higher yields with high salinity water and this could be achieved by the sprinkler, drip or the surface irrigation methods.

4. **Conjunctive use of fresh and saline waters:** There are situations where good quality water is available for irrigation but not in adequate quantities to meet the evapotranspirational needs of crops. Under these conditions, the strategies for obtaining maximum crop production could include mixing of high salinity water with good quality water to obtain irrigation water of medium salinity for use throughout the cropping season. Alternatively, good quality water could be used for irrigation at the more critical stages of growth, e.g. germination, and the saline water at the stages where the crop has relatively more tolerance. Further research is needed to define the best options considering the tolerance of crops at different growth stages, critical stages of growth vis-a-vis soil salinity, etc.

5. **Cultural practices:** Cultural practices can often be modified to reduce the hazard of high salts in the irrigation water. Similarly a modification in the method of irrigation can result in improved use of water for some crops. These aspects have been discussed earlier.

Management practices for efficient use of water with sodicity hazard

As in the case of irrigation water with a salinity hazard, appropriate management practices can often help in better and more efficient use of water with a high sodicity hazard. These practices include:

i) **Application of amendments:** Since accumulation of the sodium ion on the exchange complex is mainly responsible for poor soil physical properties, irrigation water having a sodicity hazard could be improved by increasing the soluble calcium status of the water, thereby decreasing the proportion of sodium to the divalent cations and therefore its adsorption on the soil exchange complex. Applied soluble calcium salts will also neutralize the bicarbonate and carbonate ions thereby reducing the sodicity hazard of the water. The quantity of an amendment that must be applied, the mode and frequency of application etc., are some of the practical questions. Bhumbla and Abrol (1972) recommended that for RSC values up to 2 mmol (+)/1 there was no need to apply an amendment. For higher RSC values, the required amounts of amendment should be calculated and the recommendations made accordingly. Thus the gypsum needed to decrease RSC by 1 mmol (+)/1 works out to 850 kg per hectare metre of water. Gypsum can be either incorporated in the soil or lumps of gypsum can be suitably placed in the water channel to dissolve gradually.

Sulphuric acid has also been used to amend water quality and can be applied directly to the soil or in the irrigation water. It rapidly neutralizes the sodic constituents of water or reacts with lime in the soil to produce soluble calcium. On an equivalent basis, however, the effect is nearly the same as that of gypsum. Being corrosive, handling of sulphuric acid presents problems which must be overcome through proper application techniques.

ii) **Mixing with an alternate source of water** If an alternate source of irrigation water is available, mixing the two sources may be helpful in obtaining water which is acceptable for irrigation considering its sodicity hazard. Detailed chemical analysis and the quantities in which the water is available from the two sources can help in deciding the proportions in which they need to be mixed.

iii) **Irrigating more frequently** Irrigating frequently with small quantities of water is an effective way to manage water with a sodicity hazard. Reduced permeability of the soils restricts water supply to the roots. Also applying large amounts at a time can result in surface stagnation which affects most crops adversely. Frequent irrigations could also reduce the precipitation of calcium by reaction with bicarbonates in water by keeping the soils wet. Using sprinkler irrigation with the ability to supply controlled amounts of water at a time should be considered where feasible.

iv) **Growing crops with low water requirements** hen the irrigation water tends to create a sodicity problem, it is advisable to use small quantities of water, waters with significant quantities of residual sodium carbonate (RSC) will cause a continuous increase in the exchangeable sodium status of soils and therefore the need to limit water use. Unlike saline water, where application over and above the evapotranspiration requirements is recommended, extra application of water with a sodicity hazard will further aggravate the problem. If feasible, growing crops and irrigating during periods of high evapotranspiration demands should be avoided.

v) **Growing tolerant crops** Growing crops tolerant of excess exchangeable sodium and poor soil physical conditions will help obtain better returns than if sensitive crops are grown.

Relative crop tolerance to soil sodicity

Sensitive (ESP< 20) : Gram. Soybean, Saffiower, Blach gram, Urdben, Lentil

Semi Sensitive (ESP 20-30): Linseed Garlic, Onion, Groundnut and Guar

Moderately Tolerant: (ESP 30-50) :Mustard, Wheat, Sunflower Rape seed, Sorghum, Pearl millet, Pigeon-pea,

Tolerant (ESP >50): Rice, Sesbania and Barley

vi. **Organic matter applications** Heavy dressings of organic manures, regular incorporation of crop residues, application of such organic materials as rice hulls, sawdust, sugar factory wastes, etc., have all been found useful in maintaining and improving soil physical properties and in counteracting the adverse effect of high levels of exchangeable sodium. Wherever feasible therefore, organic matter applications are especially recommended if irrigation water has a sodicity hazard.

The agronomical management practices for efficient use of saline water in agriculture

The change in quality of irrigation water is technically not feasible and not economically viable. The following measures should be adopted for the use of saline water.

(A) Selection of crops

(i) Selection of salt tolerant crops and varieties.

High tolerant : Rice, cotton, sugar beet, tobacco, date palms, ber.

Moderate tolerance : Sugarcane, wheat, guava, pomegranate, tomato, sweet, potato.

Highly sensitive : Beans, pea, grape, orange, apple, pear, carrot.

(B) Tillage

(i) Deep ploughing to break the hard pan of salts if any.

(ii) Furrow planting is best for saline conditions because the seed can be safely planted below the zone of high salt accumulation.

(C) Seed rate

(i) Poor germination, high mortality of young seedlings and poor tillering are common features when the crop is grown with poor quality water. So, higher seed rate and close spacing is advisable. An additional seed rate of 25% should be adopted.

(D) Organic and inorganic fertilizer

   (i) Dhaincha as a green manuring crop improves physical properties of the soil.

   (ii) Addition of organic manures to some extent mitigates the adverse effect of poor quality water.

   (iii) Fertilizers should be applied 1.25 - 1.5 times the normal rate of their aplication.

   (iv) Application of Zn @ 20 kg ZnSO4 ha^{-1} counteracts the negative effect of higher salinity and sodicity.

   (v) Split application of nitrogen to prevent N losses through volatilization. Correction of nutrient deficiencies by foliar application of nutrients.

(E) Irrigation

   (i) Dilution and cyclic use of good and saline waters. When good quality water is limited, it can be used as follows :

   (ii) Pre sowing and first irrigation should be with good quality water. Later saline water can be used.

   (iii) Poor quality water can be mixed with good water.

   (iv) Drip or pitcher irrigation is found suitable.

(F) Mulching

   (i) Use of mulches to reduce the requirement of water for evaporation. Use of mulches and intercultural operations reduce water requirement of crops, thus with saline water salinity develops at a relatively lesser intensity.

   Salt tolerance of different crops.

   High tolerant: Rice, cotton, sugar beet, tobacco, date palms, ber.

   Moderate tolerance: Sugarcane, wheat, guava, pomegranate, tomato, sweet potato.

   Highly Sensitive: Beans, pea, grape, orange, apple, pear, carrot.

REFERENCES

Brady, N. C. and Weil, R. R. (2002). "The Nature and Properties of Soil".13th Ed. Pearson Education, Asia.

Das, D. K. (2013). Introductory Soil Science, Kalyani Publishers, New Delhi.

Das, P. C. (2009). "Manures and Fertilizers". Kalyani Pub. 3rd Ed.

Hall Troeh, F. R. and Thompson, L. M. (2005). "Soil and Soil Fertility". Blackwell Pub.

Havlin J. L., Beaton J. D., Tisdale S. L. and Nelson W. L. (2006). Soil Fertility and Fertilizers. 7th Ed. Prentice

Jurinak, J. J. (1978). Salt-affected Soils. Department of Soil Science &Biometeorology. Utah State Univ.

Kabata Pendias, A. and Pendias, H. (1992). "Trace Elements in Soil and Plants". CRC Press.

Kannaiyan,S., Kumar, K. and Govinarajan, K. (2004) "Biofertilizer Technology". Scientific Pub.

Kanwar, J. S. (1976). "Soil Fertility: Theory and Practices". ICAR, New Delhi.

Mehra, R. K. (2017). Textbook of Soil Science, ICAR, New Delhi.

Mengel, K. and Kirkby, E. A. (1982) "Principles of Plant Nutrition". International Potash Institute, Switzerland.

Mortwedt, J. J., Suman, L. M., Cox, F. A. and Welch, R. M. (1991). Micronutrients in Agriculture. 2nd Ed. SSSA. Madisson.

Rajani, A. V. and Jadeja, A. S. (2019). Fundamental of soils and plant nutrients. Jaya Publishing House, Delhi (India).

Richards, L. A USDA Handbook No. 60. (1954). Diagnosis and improvement of Saline and Alkali Soils. Oxford & IBH.

Singh, S. S. (2011). Soil Fertility and Nutrient Management, Kalyani Publishers, New Delhi.

Somani, L. L. and Kanthaliya, P. C. (2004). Soil and Fertilizers at a glance, Agrotech Publishing Academy, Udaipur.

Stevenson, F. J. and Cole, M. A. (1999). "Cycles of Soil: Carbon, Nitrogen, Phosphorus, Sulphur, Micronutrients. John Wiley and Sons.

Yawalkar, K. S., Agarwal, J. P. and Bokde, S. (1992). "Manures and Fertilizers". 7th Ed.